"Stan Rowe . . . was a teacher, philosopher, and ecologist, and touched many Canadians . . . His writing is elegant, strengthening our understanding of the relationship of humans to nature. The essays teach but do not preach. Yet, there is a central message—the earth, our life support, is finite. We cannot circumvent this reality regardless of new technologies. If we ignore this message, we will perish."

—Nikita Lopoukhine, Chair, IUCN
Commission on World Protected Areas

"The conceptual revolution necessary if humanity is to pull out of the nose-dive we are in requires the subtle and not-so-subtle insights derived from essays in this book."

—Wes Jackson, President, The Land Institute

Earth Alive

ESSAYS ON ECOLOGY

STAN ROWE

Library and Archives Canada Cataloguing in Publication

Rowe, J. S. (John Stanley), 1918-2004.
Earth alive : essays on ecology / Stan Rowe ; edited by Don Kerr.
Includes bibliographical references.
ISBN-13: 978-1-897126-03-5
ISBN-10: 1-897126-03-4

1. Human ecology. 2. Environmentalism. I. Kerr, Don II. Title.
GF75.R69 2006 304.2'8 C2006-900925-2

Editor: Don Kerr
Cover and interior design: Ruth Linka
Cover image: Harry Savage
Author Photos: Peter Jonker

NeWest Press acknowledges the support of the Canada Council for the Arts and the Alberta Foundation for the Arts, and the Edmonton Arts Council for our publishing program. We also acknowledge the financial support of the Government of Canada through the Book Publishing Industry Development Program (BPIDP) for our publishing activities.

NeWest Press
201–8540–109 Street
Edmonton, Alberta T6G 1E6
(780) 432-9427
www.newestpress.com

NeWest Press is committed to protecting the environment and to the responsible use of natural resources. This book is printed on 100% post-consumer recycled and ancient-forest-friendly paper. For more information, please visit www.oldgrowthfree.com.

1 2 3 4 5 09 08 07 06

PRINTED AND BOUND IN CANADA

To Stan and Katherine,
the pleasures of New Denver

JOHN STANLEY ROWE, BORN JUNE 11, 1918, DIED APRIL 6, 2004

TABLE OF CONTENTS

PREFACE

I first saw Stan Rowe at a University of Saskatchewan council meeting. Someone had used a new word, ecology. The University President, a physicist, didn't know what it meant, nor did the Dean of Arts, a chemist, nor did the Dean of Agriculture, but he said one of his faculty members, Professor Rowe, might explain the term. As a poet and English professor I was skeptical of new words, but when Stan, in his polite way, defined ecology, I realized that the word opened up a world I had never imagined.

In time my response was to ask Stan to join the *Newest Review*, which I was editing in the early '70s, as environmental editor. He accepted, reviewed books, wrote articles, always with his splendid prose, a sure way to reel in an English Prof. Later, having read others of his essays, I suggested he should publish a book of essays. Stan, so clear and strong in his own views, and for whatever reason—lack of ego?—demurred. I had to convince him to do the book. Unbelievable. He said he'd do it if I helped as editor.

We spent two months as I recall, essay by essay, removing those passages Stan had recycled—when he had a good insight he'd repeat it for different audiences—then organizing the essays under topics. We often worked at the Faculty Club. Stan would arrive by car, I by bicycle, a renewable resource—until Stan felt guilty and took to walking. All that work became *Home Place: Essays on Ecology*, 1990, new edition in 2002.

When I recently wrote about my own writing I decided that editing *Home Place* was more important than any of my own work.

Four years ago Stan emailed NeWest Press to say he was planning a second book of essays, knew his contract said the Press had the right to his next book, but planned to publish with another publisher. I phoned Stan, encouraged by Ruth Linka, and he said yes to NeWest—what he needed was someone to help make him finish the manuscript. So we began again, this time long distance since Stan on his retirement had moved to New Denver—you'll read all about why in this book. He sent essays by email. I replied by mail. Every so often a great swatch of material would arrive. Then there was silence as Stan wrote something new. Then another batch arrived. I was always excited as I printed off the pieces. Then in the summer of 2004 Stan died and I had to finish the book without him, a matter of great sadness.

We had already agreed on some matters, on the title, on the essay to open the book, on the section "New Denver", on including the shorter pieces, some of them wonderful, and on including reviews. I didn't know how I'd finish the book but phoned Katherine, Stan's partner and executor, and she welcomed the book's completion. So we drove to New Denver later that same summer and Katherine invited us to the office at the back of the property (house, garden, office).

There was a great pile of manuscripts on a table but they were what I already had. Our son David knew computers and on Stan's machine we found the whole book. Stan had already made many revisions, especially paring down reviews. I'd reread all the material on the trip so I had small changes too. In ten hours over two days David and I had the book, not in any order except Stan's machine shorthand—Heavies, Shorts, Reviews. I was delighted. Stan had finished the book before his stroke.

I've had to do the organizing on my own this time, omitting some Heavies because of repetitions. We can't fine tune things now.

My own debt to Stan is simple. He changed my view of what is important. Live on Stanley in your splendid words.

—Don Kerr

INTRODUCTION

Before the word "ecology" entered popular parlance I was a student of ecology in the late '30s, studying the flowering plants and soils of the landscapes in Alberta and Saskatchewan with Dr. Ezra H. Moss and Scotty Campbell. A decade later, after the war, I resumed ecological studies on the grassy plains of Nebraska with Dr. John E.Weaver and then, encouraged by Drs. Askell and Doris Love, in the southern boreal forests of Manitoba and Saskatchewan. Employment through the years as a research ecologist and teacher of ecology, with the Canadian Forestry Service and the University of Saskatchewan, kept my eyes open to realities beyond human society. I mention these facts to explain my focus on Earth and its terrarium/aquarium parts in the following essays and reviews, for this book is about human ecology, about our relationships to the large Being that surrounds and gives life to us: Planet Earth.

All organic things are systems of smaller parts, as well as ecologically related parts of larger surrounding systems. For example, inwardly the human heart is a vital system of tissue parts, while outwardly it is an animated part of the whole body system. Again, each person is composed of organ-system parts (for breathing, digesting, circulating) while also an interacting ecological part of the enveloping Earth-system (that provides the air for breathing, the food for digesting, the water for circulating). Although we are wholes in relation to our inner organs, we are parts in relation to the outer world—parts dependent on the whole Earth for our existence.

Every living organism is autonomous to a degree, free to be itself and to do its own thing with one important limitation; namely, the responsibility to maintain the integrity of the whole of which it is a part. This is the overriding duty of cell to tissue, of tissue to organ, of organ to body, and of human body to Earth. When cells disobey their ecological responsibility within the body, the result is cancer. When people disobey their ecological responsibilities, the result is over-population and degradation of the Earth and its ecosystems.

That every human being is made up of smaller parts that work together for our health is obvious. Not so obvious is the fact that each one of us is a dependent part of a surrounding ecosystem whole, with responsibilities for maintaining its welfare. The idea of a primary obligation to maintain the integrity and health of Earth's supportive ecosystems, and the corollary that there must be limits on the freedom to do whatever we want within this larger Being, goes against tradition. Hitherto, society has placed high value on the liberty of the individual, backed by private property rights, to convert Nature's wealth to human wealth. This liberal viewpoint was serviceable in the past when the main dish on the political menu was resistance to the powers of King and State. Now it is stale, the cause of environmental dyspepsia. Ecological sense is prescribed.

Ecological sense does not come easily to us moderns and post-moderns. We have inherited a history of philosophic idealism that values ideas above the perception of Earth as revealed by the senses. The international world religions and philosophies tend to be other-worldly, more concerned with escape from material reality and with evasion of death, than with Earth-celebration. The history of science shows a fixation on the inward viewpoint that promotes the importance of physics and chemistry, with slight attention to the outward ecological viewpoint that relates us to Earth. These threads are integral parts of Western culture's paternalistic weave that highlights mind/soul over body, male over female, rationality over feelings, and sight over the other senses.

Perhaps the tide is turning. Philosophy since Goethe has taken more seriously the reality of the human body-mind and "the world as flesh." Earthly religions and Aboriginal animisms that perceive life in all things are finding their intellectual proponents and supporters. Science, whose technology is the cause of environmental woes, must perforce stop playing power games with atoms and molecules, turning its attention to solving the problems heaped on Earth. Feminists are displaying to public view the lop-sided and disfiguring results of a culture where male rationality, sight and soul rule to the disadvantage of the body, the female and the sensual world. The tout ensemble urges humanity to get real, to leave cloud cuckoo-land and come down to Earth.

Those who seek prime values in the living Earth, who perceive it as the chief reality for humans, may be dismissed as "materialists." This term is rightly used as a put-down when applied to those who see in the material parts of Earth, Moon and Mars only their utility for the human race: celestial bodies with land for mining, trees for logging, grizzly bears for shooting, and rivers for generating electricity. But philosophic materialism/naturalism need not be selfish, nor unaware of its unsubstantial counterparts: the parallel realities of mind, intuition and intelligence. It trusts our senses to glimpse reality, which is revealed as having a material basis. Thus the material is elevated and perceived as inherently valuable, which is surely good compared to its deadly opposite: the desire to own, use, and consume, solely in the interests of the individual or society. Heaven may need its supernaturalists who discount the material reality of their lives, but Earth needs its ardent naturalists who glory in their material existence.

1. NEW DENVER

HOW TO GET TO NEW DENVER

How come a prairie boy, child of the grasslands, ends up in the Interior Wet Belt of British Columbia, surrounded by tall Douglas fir, cedar, and hemlock? I could blame it on my mother and father.

Arthur Herbert Rowe came to Alberta from Ontario in 1912 as a minister of the Methodist Church that in 1925, by union with Congregationalists and most Presbyterians, morphed into the ecumenical United Church of Canada. His anti-war stance was strengthened by the effects of the First World War on two of his brothers who survived military service overseas but were deeply damaged mentally. My mother lost her only brother in the same war when his troopship was torpedoed. Pacifism was the family philosophy, and from time to time we entertained at our home such then-famous peace proponents as Carlyle King and A.J. Muste.

When World War II began in late 1939, I was a sophomore at the University of Alberta. A small group of pacifist students, members of the Fellowship of Reconciliation, met regularly to discuss the war and provide mutual support. They helped me through the mental turmoil of preparing for the inevitable conscription "call-up" that came in my third university year. I responded by declaring myself a conscientious objector, and was ordered to appear before a military tribunal who had me tagged as a draft-dodger. One of the three suggested that if I enlisted without further fuss I would soon be an officer, thus avoiding the dangers to which less educated troops were exposed. When I argued the point, protesting that pacifism was a philosophy of non-resistance and not a means of escaping danger, they shook their heads. Their definition of a "pacifist" was a bible-carrying member of minority anti-war religious sects: the Hutterites, Doukhobors, Mennonites, Seventh Day Adventists, Pentecostal, Jehovah's Witnesses, and the quiet Quakers. My affiliation was United Church and, like members of the Anglican Church, did we not sing "Onward Christian Soldiers"? That was true. The main stream churches supported the war and condemned clergy who preached otherwise. Their divinity students were exempt from conscription, presumably with the expectation that after ordination they would overlook the teachings of their Founder and bolster morale for the war effort.

No action against me was taken until I graduated in the spring of 1941 when another call-up notice arrived with the warning that failure to comply meant a jail sentence. Again I sent back a refusal and waited. A sympathetic professor in the university's Department of Agriculture gave me part-time employment weeding cereal plots. A couple of months passed, then one late-summer day two Mounties drove up in a police car. "They've come for me," I said to my surprised co-workers. The police were not unfriendly and on the way to the station they read me my rights and suggested I say as little as possible—doubtless to cut short the nervous chatter. From the station I phoned my landlady who had been warned that sooner or later I would not turn up for dinner. She notified my far-away parents, and they suffered more mental anguish than I. "Taken into custody" proved less traumatic than imagined.

The next stop was the provincial jail at Fort Saskatchewan, east of Edmonton, where I was relieved of my clothes and meager possessions, and given the traditional grey prison suit. I had my own barred cell, about ten by twelve feet, with bed, basin and flush toilet. An outside-facing window was paned with rippled glass, but a couple of small holes at the bottom of the window frame let me peer out at the North Saskatchewan River valley, beautiful in its early autumn colouration with a few dark green spruce towering over the yellowing poplars. Once I saw a solitary dark figure walking the riverbank who I guessed to be Dr. Turner, a local amateur botanist of whom I had heard at the university. Oh, to be sauntering and botanizing with him! The close confinement, the slamming echo of the steel cell doors, the occasional rattling of the bars and yelling by the inmates, suggested zoos at feeding time. The jail food strengthened the comparison. A concession to health and well-being was a weekly ration of roll-your-own tobacco.

Fortunately my sentence included the words "with hard labour" which meant that the worldly-wise Warden could send me outdoors to help trim, dig, and transplant amongst the flower beds and lawns of the prison grounds, separated from the tougher element who got their daily exercise in a guarded enclosure. This good fortune turned out to be educational. The head gardener was a fiery little Scotsman, a friend of Tim Buck who for many years was secretary of the Communist Party of Canada. He and Buck had risked their lives in 1925, attempting to radicalize the coal-miners in Drumheller. He told stories of clandestine meetings at night with fellow conspirators while the Mounties tried to hunt them down. My working mate was a handsome young felon who believed, ahead of his time, that day labour was for the weak-minded. A "keister cracker" by trade he admitted having erred on his last nocturnal outing when, after gaining entry to a second-story office via a handy telephone pole and successfully opening the door of a steel safe, he had found along with considerable cash a bottle of Scotch whisky. Hours later, his head still foggy, it occurred to him that escape was perhaps his next priority, and in broad daylight he shinnied down the pole into the waiting arms of the law. Sharing "hard labour" with me for a month convinced him, against his earlier principles, that an argument could be made for using one's muscles and not just one's head. Prison reform works in odd ways.

Meanwhile letters were pouring in to the government protesting that the son of Rev. A.H. Rowe was a bona fide conscientious objector. Supportive letters came to me from my family and other anti-war people, including one from J.S. Woodsworth, founder of the CCF (now the NDP) and himself a declared pacifist, saying in effect, "Good for you! Hang in there; we're working to get you out!" And so, after a couple of months of internship, a guard took me down to the office of the Warden who gave me back my clothes and turned me loose. What a wonderful feeling! Free again!

I went home for a few weeks, relieved to be able to wander about at will, and then received a letter telling me to report for Alternative Service at Calgary. The federal government did not quite know what to do with conscientious objectors, so asked the BC Forest Service to provide employment in the woods. About thirty of us young men—mostly of the Mennonite, Hutterite, Pentecostal, and Seventh Day Adventist faiths—were bussed to a logging camp near Radium Hot Springs. Our first job was cutting down mature lodgepole pine infested with bark-beetles, a salvage operation justified by the belief that removing infested trees would keep the bugs from spreading. Now it is known that the only effective control is winter cold—and that we experienced. One of the Mennonite boys found in my *Pocket Book of Verse* the poem from Shakespeare's *As You Like It* whose lines, particularly the last one, he thought appropriate for our situation:

Who doth ambition shun
And loves to live i' the sun,
Seeking the food he eats
And pleased with what he gets,
Come hither! come hither! come hither!
Here shall he see
No enemy
But winter and rough weather.

From the interior of British Columbia we were moved to the milder climate of Vancouver Island and put to work planting Douglas fir seedlings on "the Courtenay Burn"—a gigantic clear-cut area that had been overrun by fire. Here the story was told that before the war so many unemployed men from Vancouver had been conscripted to fight the fire that appropriate tools were soon in short supply. When additional long-handled shovels were requested from head office, a lack of trust in the assiduity of the fire-crews was indicated by this terse response: "No more shovels; tell the men to lean on each other." I returned to the plantation site in 1998 and found that the Douglas fir seedlings, planted as we thought in the interests of restoring a beautiful forest, had grown to a "merchantable" size and were being logged by Crown-Zellerbach.

Pearl Harbor was bombed late in 1941 and a few months later the federal government, at the instigation of racist BC politicians, used the War Measures Act to order the removal of all Japanese Canadians living within one hundred miles of the Pacific coast. Although no Japanese Canadian was ever charged with disloyalty to Canada, more than twenty-thousand men, women and children—three quarters of them Canadian nationals—were removed from their homes in 1942 and sent eastward, mostly to detention camps in the interior of the province. That summer the Japanese military released incendiary balloons, a few of which floated over the Pacific and landed here and there on the BC and Alaska coasts. A forest-fire scare resulted and our "alternative service" was quickly switched from planting trees to cutting down snags—the dry, flammable ghosts of trees standing here and there on logged and burned sites. A Norwegian woodsman taught us the arts of using the double-bitted felling-axe, the two-man cross-cut saw, and the spring-board that elevates sawyers above the butt-swell of old firs and cedars.

The provincial government refused to fund education for the Japanese-Canadian children in the evacuation centres, while the federal government only undertook responsibility for providing the lower or public school grades. Church groups stepped in to fill the educational gap. Gwen Suttie, home from Japan at the time of Pearl Harbor, was asked by the Women's Missionary Society of the United Church to set up a kindergarten and a high-school in New Denver. At the same time a churchman in Vancouver got permission from the federal government "to borrow" from Alternative Service the small number of conscientious objectors who had university training. To my surprise I was called to the boss's office—at the time in a forest camp near Courtenay on Vancouver

Island—and told I had been listed as a potential high-school teacher. Would I consent to be a teacher in the BC interior? You bet!

Gwen Suttie had been given a list of prospects, and my credentials caught her eye: a science major on his way to Harvard but for the war, and son of a minister of the church. Fortunately, she discounted the rumour that preacher's kids are notoriously wayward (at the University of Alberta, a large group of students with clergy parents had formed the PKA—the Preachers' Kids Association—soon dubbed the Promiscuous Kissers Association). Lucky me, to be invited to New Denver on a worthy project: teaching Grades 9 to 12 science and maths, using correspondence courses. Our school, the little Turner United Church, was divided by curtains into four sections, one for each of the four grades. An equal number of teachers circulated from class to class and, by the end of the day, each had gone once around the inside of the school. The Nisei, second-generation Japanese Canadians, were a friendly and likeable group, all glad to be back in school with their peers. Generally they did well academically and we had no discipline problems. The years 1943–45 and thereafter were hard ones for the dispossessed adults, less so for the resilient young people.

Chance brought me the first time to New Denver. The second time, after retirement in 1990, I chose it. The Turner Church/School still stands. Nearby is the house I shared with social-science teacher Mildred Fahrni, with musician-English teacher Helen Lawson, with kindergarten teacher Ella Lediard. Now and again Nisei visitors from Canada and the USA come back to visit and say hello. Sixty years have seen little change in our village.

VILLAGE LIVING IN CITY AMBIENCE

Like most young people from small towns I migrated to the city after high school graduation, and thereafter spent the greater part of my time in the urban centres of Canada and the USA. Likewise in travels abroad the destinations were always cities and, secondarily, their hinterlands. But in retirement I have cast off city ways, moved out, and am ensconced now in the boondocks. Talking to friends from over yonder where the traffic hum never ceases I sometimes detect mild puzzlement and an unasked question: "Why willingly forsake the urban milieu?" It's a variant of the old post-war song, "How you gonna keep 'em down on the farm, after they've seen Paree?" reflecting a ten-thousand-year-old presumption that the city is the only place to be.

Further, whatever the reasons for rusticating, why retire of all places to New Denver?—a village with a population of about 650 plus an equal number of cats and dogs, tucked away in the mountains of southeast British Columbia 100 kilometres from the nearest airport at Castlegar and the same distance from the nearest "shopping centre": the hippy city of Nelson, population 12,000.

First the city question. It is true that many people, perhaps most, are unwilling to relinquish urban life once it has been experienced. Attuned to the city's fast pace and stimulation, to its crowds and traffic, to its mechanical day-and-night sights and sounds, they feel uncomfortable except when surrounded by human artifacts in the built environment. On vacation, they bring the city with them to tame the wild campground, looking askance at the uninhabited surroundings. Visitors to New Denver have complained that the nearby mountains are foreboding and induce mild claustrophobia, that the lake is dangerously tempestuous and, when calm, too cold for comfortable swimming, that the forests at the village edge are dark and lonely, and that bears of questionable temperament wander through the streets after dark. They feel more at home and safer where the media daily report traffic smash-ups, muggings, rapes, and occasional murders.

Others, while condemning the city for its artificiality and crime, find adequate compensation for these negatives in the panoply of cultural activities that only large concentrated populations can support. Where but in the city can one find the symphony orchestra, the sports arena, the opera company, the art gallery adequate in size to attract the travelling international show? Ah, but these are entertainment for a mostly passive public. Small communities offer the opportunity for everyone to participate actively in music, sports, the arts. In the city's large talent pool only professionals can be bullfrogs, but in rural areas where talent pools are puddles, all are encouraged to join the amphibian chorus, giving voice to their amateur interests without stigma or shame. Greenhorns can participate in almost any activity. Want to be an alto though you haven't sung since high school? Welcome to the Valhalla Choral Society!!

The community spirit is strong. An inventory estimates seventy different kinds of organizations in and around New Denver, catering to every taste. A well-equipped school with gymnasium is open for public use after classes. An abandoned church, Knox Hall, is a popular meeting place with a basement library that on request brings books-on-loan from the city. Music is performed with much pleasure by a community choir that presents public concerts in spring and mid-winter. A school band is open to oldsters, and a twenty-five-member orchestra with several members playing at professional levels performs the simple classics. For those interested in the arts, classes are available from time to time in painting, sculpture, pottery, etc. A community Gallery handles plays, cabarets, musical evenings, and the seasonal displays of local artists. A number of published writers live near the village and book clubs flourish. If this sounds too much like diluted "city culture," like a pale version of the virtual reality lived by urban folk, at least we in the village know that the Real Thing still exists just beyond our doors.

Agreed that the equation, Cities = Civilization, is valid for ardent humanists. Cities display the human monoculture in its purest form, much as a formicarium displays ants. But for some of us the questionable duty of attending to human society as Earth's central reality, to the exclusion of the non-human majority, is too limited. My biological excuse, based on observed behavior, is that the primate *Homo sapiens* is semi-gregarious, not fully gregarious like termites and yellow-jacket wasps. Our evolutionary history is that of wanderers in the wild, few in number and sometimes solitary—a state that everyone tries to recapture from time to time. Most people seek to avoid members of their own species now and again, intrigued with interesting alternatives—natural landscapes, flowers, trees, birds, butterflies. And who has not longed to escape the works of man: suburban housing, rectangular lawns, glassy office buildings, paved streets, and cars, cars, cars. One certain answer is: those with naturalists' leanings. To paraphrase the poet, it is not that we love people less but that we love Nature more.

To live without easy access to the non-domesticated world seems lamentable to me. This judgment is based on comparison of two modes of living. Much of my adult life has been divided between the wild and the city in the approximate ratio of one to three. Employed as a forester and field ecologist, my summer months were usually spent in remnant wildernesses: native grasslands and forests in the south, peatlands and tundra in the north. These spring-to-fall experiences were an unfailing delight. Breathing the crisp air, tramping through landscapes molded by the last great ice-sheet, companioned by lakes and clear streams, surrounded by beauteous plants and animals admirably fitted to their ecosystems, my mental equilibrium was restored. Looking back, it seems to me that whatever sanity possessed me in the turbulent city was a carry-over from the unconfused non-city. The untamed world is surely a source of healing power. Wild Nature revitalized me in the original sense of the word "re-creation." To live surrounded by it again, is good.

Understandably, people are drawn to the city because that is where the jobs are. The opportunities for gainful employment—even if it is housekeeping, childcare, or serving up Big Macs—go far beyond what the backwoods has to offer. Some are drawn to the city because relatives live there, to be close to their families or to their minority religious group. Educational facilities, schools and libraries, surpass those of the backcountry. Perhaps too a "city type" exists, a class of people genetically suited to the turmoil of cities, those who in the words of Eisenberg are pre-adapted to the urban life, taking to its abiotic technology, its machines, its violent traffic, its thundering herds of buses, taxis, and humans, like foraging ancestors took to the thundering animal herds of the Serengeti. I argue for rural living as a counterbalance to the present dangerous trend that, if continued, will find more and more people in mega-cities, more and more separated from an important part of their Earth-aware heritage.

Now for the second question: Why retire to New Denver after long years spent on the wheat-covered plains? Any who have grown up in the grassland region within sight of the cordillera will understand. The mountains on the western skyline were magnets for holiday time. Pine forests were there, cold clear lakes, high rock faces, waterfalls. I loved the prairie grasslands but the contrast of the nearby mountains drew me too. Memorable experiences in my teens were camping trips to the nearby Rocky Mountains: to the Livingstone Range (reputed locale of a fabulous lost mine), to Jumping Pound where the aborigines lured bison to their doom, to lovely Waterton Lakes where mountains rose abruptly from the plains with no intervening foothills.

A few years later I came to New Denver by chance, serendipitously selected to teach high school sciences and mathematics to young Japanese-Canadian evacuees from the West Coast during the war years, 1943–45. Here is superb landscape beauty: cloud-draped mountains thickly forested with many evergreen species, a mild rainy climate, spring arriving in March, golden autumns, cherry, apple and pear trees. For a few years during non-teaching hours I hiked the old mining trails, climbed peaks and paid my respects to mountain goats and grizzly bears, watched birds on the lake, identified the wild plants—at least enough to make me feel at home. And so it seemed right, in making the last move of my life, to forsake the prairie cities in favour of New Denver on the east shore of Slocan Lake—"the Lucerne of the Selkirk Mountains."

The village is sited at the mouth of Carpenter Creek on a small delta, formed during deglaciation in a series of flat steps as lake levels dropped and boulder-tumbling floods intermittently roared out of the eastern mountains. The excessively stony substrate has so far prevented installation of a sewer system (every house has its septic tank and disposal field under grass and garden in the back yard), thus preventing high-density housing. To the frustration of realtors, no condominiums have sprouted along the lakeshore and so the "eyes to acres" ratio in the village remains gratifyingly low.

The valley occupied by Slocan Lake and the short Slocan River that drains it southward into the Kootenay-Columbia system is home to about 5600 people in a scattering of unincorporated communities, quasi-communes, hay farms, stump ranches, and three small villages of which New Denver is the largest. About half the valley population is in the labour force, mostly employed in the retail trade, in providing accommodation for travellers and tourists, in social services (especially health and education), construction, small manufacturing, logging and forestry. One quarter of the valley's economic income is contributed by those who are financially independent—the retired and semi-retired. Agriculture is minimal, mostly the growing of hay, because of the proximity to the more productive Okanagan Valley. Nevertheless many small organic farmers produce vegetables, fruit, milk and meat for local consumption.

Like many other rural areas, the Slocan Valley is characterized by high unemployment and uncertainty about job retention in the resource industries, chiefly forestry now that the mining-boom is long gone. Mechanization and an erratic market combine to keep employment low in the woods and at the mills. On the plus side small business and home-based cottage industries are ubiquitous, reflecting the flexibility of a do-it-yourself work force whose employment is partly outside the area, thanks to electronics and a network of surfaced roads. Women tend to be the organizers and leaders of communal enterprises.

The ingenuity of the population is indicated by an official estimate that places the proportion of regional income from underground activities at 5 to 10 per cent, roughly five to ten million dollars annually (the unofficial estimate, doubtless truer, doubles that amount). Bartering, wildcrafting, harvesting of non-timber products, subsistence activities (hunting, fishing, gardening) including the growing and selling of illegal substances, are relatively common in the Slocan Valley. Rarely a fortnight passes without a report that the police have detected yet another grower or possessor of marijuana. Perhaps the ingenuity of the locals results partly from culture hybridization, for we are a healthy mix.

The Aboriginal people in what is now south-central and south-eastern British Columbia were migratory, in the winter moving southward to warmer climes where they maintained their relatively permanent camps. When the Canada-USA border was drawn along the 49th parallel the Sinixt were cut off from their traditional summer foraging grounds: the lakes, streams and lands of the Kootenay region. Consequently year after year their numbers in Canada dwindled, and the government was pleased to declare them extinct. That official pronouncement, based on hope rather than fact, was premature. Today, here as elsewhere, native voices are demanding justice from the non-migratory majority who put their trust in private property with its borders, boundaries, and fences.

The first invading privateers were not the usual missionaries but miners. In the 1800s mining was big in the North American west, advancing northward from gold in California to silver in Nevada to copper, silver, gold and lead in Montana and neighbouring Idaho. The next jump late in the century was across the border into the Kootenays where outcropping veins of silver-lead-zinc were discovered within and adjacent to the Slocan Valley. Names such as "New Denver" and "Mount Idaho" (the nearby fire-lookout mountain) bespeak homesick migrants from south of 49. When the mining boom had run its predictably short course, forestry stepped in to play the theme that Canadians have perfected: export, export, export.

World War II brought to the interior of BC an influx of thousands of Canadians of Japanese ancestry, cruelly uprooted from the West Coast and detained in a scattering of temporary shack settlements called "relocation centres." Preceding them, Doukhobor migrants from the Prairies had arrived in the Kootenays in the early 1900s, some settling in the south end of the Slocan Valley; they were generous in providing home-grown vegetables until the internees were able to grow their own. New Denver still maintains its residential section of small houses and fine gardens in "The Orchard," where land was cleared of fruit trees to build make-shift housing for several hundred Japanese Canadian families. At the end of the war most moved eastward, though a minority stayed to the benefit of the village. Several of the old Orchard buildings now constitute a museum—the Nikkei Centre—that tells the story of the wartime evacuation and something of the injustices suffered by a patient people.

Doukhobor members of the radical "Sons of Freedom" sect were in turn persecuted during the 1950s when their school-age children were abducted to New Denver and, behind barbed wire to exclude their anxious mothers, educated to be good Canadians. Although intermarriage is common now, the old Doukhobor traditions of pacifism, vegetarianism and full-volume *a cappella* singing persist.

Next in the '70s came the Vietnam war refugees from the western States: worried parents with sons soon to be of military service age, as well as college-educated young "back-to-the-landers" fleeing the draft. These counter-culture migrants, most of whom are now in their fifties and early sixties, are the backbone of current resistance to forest clear-cutting and other threats to the integrity of mountain watersheds, for New Denver and neighbouring Silverton still boast good unchlorinated drinking water. With the Californians came Mexican food, and a restaurant in nearby Rosebery carries on the tradition.

The most recent immigrants to this fertile mix are the retired, those looking for landscape beauty, peace, quiet, and of course the comforts that cheap electricity makes available. Although their attitudes toward forest and water issues are ambivalent, a minority tend toward values of the New-Age set rather than toward those of the conservative (now misnamed "Reform") peaked-cap set, the latter sometimes inappropriately referred to as "rednecks" and "yahoos." Au contraire, these are the willing workers, the useful men laden with tools and machines, the uncitified hewers of wood and drawers of water who, by and large, fit themselves unquestioningly to the economic system. They maintain the roads, make up the fire brigades, and keep the villages tidy. They include the membership of unions essential to the "resource industries," the proletarians in whom Karl Marx astonishingly placed high hopes for social change.

It seems to me that the primary value of rural living, for those offered the opportunity, is the possibility (foreclosed or drastically limited in cities) of coming to terms, individually and socially, with this Earth that briefly we occupy. The phrase "coming to terms with Earth" may sound strange, implying as it does the presence of high values in the surrounding non-human matrix. The doctrine is unorthodox as are its moral corollaries, for they open a different perspective on the ancient question as to purpose in our lives. Orthodoxy has taught altruism as among our highest values: the Golden Rule, yes; the Good Samaritan, yes. But little has been planted in our childish standards of ethical behavior beyond duties and responsibilities to family, friends, nation and international community. Our horizon of care has been restricted to God's chosen species: the ensouled one, the intelligent self-conscious one. The value of this humanistic creed is limited by its narrowness. When expressed as species selfishness, its outcome debases maternal Earth by attacking the health of her lands, seas, air and other organic creatures.

The creed of chauvinistic humanism is as strong in the country as in the city, preached in the churches, taught at the core of the school curriculum. It is the taken-for-granted background of news and stories in the media as well as in everyday conversations. Perhaps the mental promptings to question its truth and expose its falsehood come a little easier to minds in small communities where "Nature" still retains something of its undomesticated meaning. Here the direct dependence of people on a clean and healthy environment, on untreated potable water, on garden produce, on cows and goats, on fruit trees, can be appreciated in ways impossible in the packaged city. Yet even in rural living, with closer exposure to Earth's natural processes, the importance of fitting human enterprises to the enveloping ecoregion, rather than the reverse, continues to be low in priority. The primary importance of our ecological dependencies are obscured in both city and country by technologies, powered by electricity and hydrocarbons, that promote a specious view of our separateness from Earth. Money buys what we need—the sources of things we consume are mostly invisible.

What residents of New Denver refer to as their special "quality of life" is usually expressed in terms of beauty in the natural environment: the white snowfields of the mountain tops, the pollution-free air (except when wood stoves are fired up in cold weather), the transparency of the lake, the non-chlorinated drinking water, the biocide-free vegetables and meat, the as-yet-unlogged mountain-side forests, the deer that at night browse unprotected gardens, reclaiming the delta that was once their winter habitat. All these and more are recognized by residents and visitors as values, especially their utilitarian merit for drawing in tourists, but rarely as fundamental values with sublime resonances. Citizens have blockaded roads to defend ancient forests and critical watersheds against clear-cut skinning of the land, but the primary motivation for protection is not the clarion call: Earth First! People's needs and wants come first, mere environment is a distant second.

The result is the winning of a few small battles but a gradual loss of the industrial-commercial war on Nature that has already left marks of its ravages on much of North America. Slowly and incrementally the domestication and simplification of landscapes proceeds, not yet effected totally in the Slocan Valley but doubtless soon to come. Call it "vilification" in the old sense of the word, a degrading infection that spreads outward from burgeoning cities to all their hinterlands.

The values of the industrial city are all-powerful; inexorably they affect both their near and far surroundings. Money is the driving force and the signs are everywhere. The Ministry of Logging (aka the Ministry of Forests) announces that resources of the valley "must be shared," meaning that an unsustainable annual allowable cut will be maintained in the interests of government revenue, wages for labour union members, and dividends for stock-holders wherever in the urban world they chance to dwell. The first clear-cut blocks are appearing on the valley mountainsides, providing trees to be turned into two-by-fours for export by the mill at the south end of Slocan Lake. Roads are being widened and straightened by blasting and scarring the lakeside cliffs, facilitating faster travel by the auto-centric society between urban centres and the valley. During shopping hours, half-ton and three-quarter-ton trucks, parked but exhaling poisonous fumes, often outnumber people on New Denver streets. Wood-chip trucks as big as boxcars thunder through the village at all hours day and night, feeding the pulp mill at Castlegar. The provincial Department of Health is kept busy issuing boil-water advisories to various valley communities as enteric infections such as beaver-fever and cryptosporidiosis become more prevalent due to the effects of poor land management on domestic water supplies. Lake fertilization is proposed to increase fish stocks depleted by touristy over-fishing. Noise, especially from internal combustion engines (chiefly lawnmowers, weed whips, motor boats and helicopters), increases year by year—an index of rising energy use approved by commerce, as more and more humans armed with more and more power throw their cumulative weight around.

The planning, the decisions, the lifestyles of the city determine what happens to the nation and to Earth. Therefore the chief "city problem" is ecological, having to do with the impact of cities on the world outside their borders. Does a healthy symbiotic relationship exist between city and hinterland? Most urban studies tend to ignore these outward ecological relationships, concentrating instead on such internal problems as revitalizing downtown commercial centres, facilitating traffic flow, and planning suburban subdivisions and city parks. Yet the fact that cities do not produce what their inhabitants most need—food, water, building materials—means that productive hinterlands are essential for their survival. Further, cities must export their garbage and wastes. Both urban and rural folk need to understand their close economic and social connections, mediated by air, soil, water, and those parts of Earth named "resources."

If the way people live and interact with planet Earth is to be reformed in the light of ecological knowledge, fundamental changes in attitude and behavior will be necessary, especially in cities. Urban people, increasingly in the majority, will then understand and appreciate the reality of what lies outside their artificial, built environment. The revolution will not come by such useful but small acts as blue-box recycling, nor by following the rules for "a simple life" that recently appeared in one of the popular environmental/ wildlife magazines. Here are the "Twelve Guides to Conduct" that, we are told, will make a difference and help "save the world":

1. Avoid shopping (compulsive shopping, recreational shopping).
2. Leave the car parked and walk.
3. Live in a nice neighbourhood (one where you don't need to use a car to go to work or do your chores).
4. Get rid of your lawn (plant perennial shrubs and flowers, or a garden).
5. Cut down on your laundry.
6. Block junk mail.
7. Turn off the TV.
8. Communicate by e-mail (saves paper).
9. Don't use a cellular phone.
10. Drink water (not store-bought beverages).
11. Patronize your local library.
12. Limit the size of your family.

Several of these "commandments" point in the right direction: walk, don't drive; get rid of your lawn; turn off your TV; and limit the size of your family. In villages such as New Denver, few sacrifices would be required to follow all twelve, though truth to tell most of the locals are just as hooked on TV-watching, shopping, driving, and lawn manicuring as the average fastidious suburbanite. The fact that such guides to behavior are proposed as fundamental challenges to urban dwellers highlights the difficulty of living responsibly in the city. Even if everyone were able to act affirmatively on all twelve, humanity would still be a very long way from "saving the world" as claimed by the article.

The health of the planet is not endangered by the small stuff. The major peril for Earth is high-energy industrial activity of one kind or another paired with population growth. These two are the heart and guts of the city, making it restlessly tick, increasing its gargantuan appetite and its resulting wastes, eating up and polluting the world. The marvel is that so few thinkers are critical of cities as cities. The urbanized landscape is simply accepted without question as one of the most natural and admirable of human achievements. Some may admit that most cities are deficient in amenities such as exterior green belts and interior gardens, but the solution of "environmental issues" is not pressing. They cost money and can wait.

In Beckett's play, *Waiting for Godot*, written in 1953, Vladimir speaks for all who trust that salvation will come their way simply by waiting: "In this immense confusion one thing alone is clear," he says, "we are waiting for Godot to come." In today's immense confusion, many continue to sit on their hands awaiting the arrival of their personal Saviour and the miraculous birth of a New Society. They believe, and have the Gurus to prove it, that divine intervention will inevitably shape this twenty-first century. Unfortunately, paradise on Earth will not come about through faith, hope, charity, meditation and prayer, useful though these may be in aiding escape from imprisonment in the cell of selfness. Nor will activists, whether religious or irreligious, effect the fundamental reorientation needed by instituting their social-political programs. Worthwhile changes in the human condition will come through the realistic thoughts and actions of people who begin to see, however dimly, that centreing the universe on themselves and the human race is both foolish and arrogant.

Where will such wisdom arise? I believe that more barriers to illumination exist in the city than in the rural setting where one can step outside, take a deep breath, and drink in the beauty of the Earth. Admiring the sky, clouds, lake, forests, and mountain peaks, feeling a part of it all, confirms an essential truth: We are Earthlings first, humans second.

Praise to people and societies who aim to preserve large parks and wildernesses for their intrinsic values, with the side benefit that the general public, especially children, may now and then get out of the city and into the wild.

As Byron wrote in one of the verses in his poem *Childe Harold*:

There is a pleasure in the pathless woods,
There is a rapture on the lonely shore,
There is society where none intrudes
By the deep Sea, and music in its roar:
I love not Man the less, but Nature more,
From these our interviews, in which I steal
From all I may be, or have been before,

To mingle with the Universe, and feel
What I can ne'er express, yet cannot all conceal.
And in New Denver—this is no Big Deal.

ODE TO NEW DENVER

Fair Village! Girt with many a mountain chain,
Wherein we worthies live, with those less sane.
Here by the sparkling waters of the Lake
On whose east shore a rushing creek by name
Of "Carpenter" (the man whose silver stake
Brought to the site a certain tarnished fame)
Has built a terraced delta with steep banks
Whose stony soils give gardeners the fits
And though exactly right for septic tanks
Precludes the building of the Hotel Ritz,
Condos, and other pricey tourist hits.
What stories can ye tell? What rare tid-bits?

First came the salmon swimming up to spawn,
Then came the Sinixt, sheltered by wigwam
(Until the icy winds of winter blew
When off to warmer southern climes they flew
Smarter in this, 'tis clear, than me and you).

Then miners found the mountains streaked with lead
And soon the word down to Seattle spread.
Bringing the rush that made Sandon a city,
Three Forks, and even Slocan, more's the pity.
New Denver was to be the crowning glory
But boom turned into bust—the mining story.
All that remained were forests by the Lake
To fatten corporations on the make.

Ah, hoped for city that was not to be,
Yet were ye saved, time after time by fate.
By homeless Isei and Nisei
The hapless victims of West Coastern hate.

How fortunate, then, to be the chosen place
Selected for the "Sons of Freedom" youth,

Wisely removed from their parental space
For education in a barbed-wire booth.

School for delinquent boys came next, add on
Escapees from the war in Vietnam,
Fleeing from California and Oregon
To thwart conscription plans of Uncle Sam.

Last came the noble list of those retired,
The grey of head, sagacious, much admired.
Would that the young could imitate this version,
Acclaimed in the International Year of the Older Person!

The history all told, what of today?
Will little old New Denver pass away?
Oh say not so, ye Goddesses, forbid!
And on such dreadful prospects put the lid!
New Denver is too gorgeous to expire,
Going the way of Nineva and Tyre.

For lo, June cherries ripen on the tree,
Each with a worm within that's hard to see.
Here shines the sun on many a summer day
Excepting when John Anderson cuts his hay
For then the rain descends most mightily
And tourists pack wet tents and homeward flee
E'en though 'tis Supernatural BC.

Autumn's the time of pears and pickled dills
While marijuana ripens in the hills.
Then winter, when the bears all hibernate
'Tis safe at night to go out on a date.
Soon it is spring, oh joy, to burn the leaves,
What fun to see asthmatics get the heaves!

And so Earth's season's move, bringing rare gifts,
Inspiring Song, and Dance—and Essayists.
Sculpting and Painting, Quilting and other arts,
Communal goings-on that make us parts
Of this our local "Arts and Culture Week."

Praise Silverton, the Gallery, and speak
Well of the Slocan where, as this doggerel showed,
The debt that I to New Denver have owed.

A CAUTIONARY FABLE

A mountain village was in deep economic trouble. The resource industry on which it depended was slumping. Many were the contradictory notions proposed by various groups to reverse the downward trend, and on everyone's lips the question was the same: Who's to blame for the failing Frogging Industry?

The lure of mining had first attracted the local population. Rich ore-bodies, there for the taking, outcropped on the surface of the mountain ranges, and export of the easy-to-mine minerals made many investors wealthy in faraway cities.

When the ore-bodies were exhausted, the working people turned to the forests for their livelihood. Logging provided wood for the mills to process and export as lumber and pulp. Forest landscapes were skinned of their trees and many watersheds ruined. Once again, in distant cities, millionaires sprouted. Eventually, as with the mining boom, the clear-cutting boom also ended and hard times returned to the village.

What to do? Export fruit, vegetables, and grain crops? No, the locals could not compete with neighbouring valleys more favoured climatically. Export water? The quality had deteriorated because of mine wastes and erosion caused by logging. Also, other depressed communities had jumped on the waterwagon first and the market was flooded with "ExtraVirgin Spring Water" and "Glacial Fresh Ice-Cubes." Export clean air? City manufacturers of canned oxygen already had the Japanese and Chinese market sewed up. Then one day an imaginative citizen hit upon another exportable item: frogs' legs!

By good fortune the village was beside a lake on whose margins lived in teeming numbers the Slocan Three-Striped Frog, *Rana slocana trilineana*. By happy chance the originator of the idea of turning frogs' legs into cash had taken a government-sponsored course—"How to Be an Entrepreneur"—where the gospels of production and marketing were preached. "Everyone can be a seller," said the expert lecturer. The implication that everyone must also be a buyer and super-consumer was not discussed.

"What can be sold?" mused the entrepreneurial graduate, glancing at her framed Diploma above the kitchen counter. Her son had brought up from the lake a couple of plump frogs in a glass jar, and as she turned the pages of her cook book in preparation for dinner she had a sudden epiphany: frogs' legs! In such intuitive flashes are great industries born.

The proposal developed rapidly thanks to a leg-up from the Small Business Bureau. Several trips to Quebec and France nailed down a steady and dependable market and the export business flourished. Soon everyone in the village had a good job, either in the canning factory or as a frog "harvester"—a popular term suggested by the Wildlife Branch that for years had encouraged thinning the ranks of cougars, elk, bear, and other large mammals by shooting holes in them from a safe distance. "Of course," said the wildlife experts, "civilized people don't kill wild animals for fun; they *harvest* the excess, like, you know, farmers cutting hay."

The supply of frogs seemed inexhaustible and soon the more inventive of the "harvesters" began to tinker with machines to collect frogs more efficiently. That this reduced the number of jobs was not considered important. "Technology makes jobs, it does not eliminate jobs," said the inventors as they turned out ever-bigger and more sophisticated "Rana Reapers" and "Froggie Bunchers" to scour the shorelines at speedier rates. The government encouraged mechanization with hefty tax write-offs keyed to the size and expense of the equipment. As people were displaced by the machines and unemployment grew, the government mounted a public information campaign whose centrepiece was the promise: "We will create more jobs."

Then disaster struck. A money crisis in France was followed by collapse of the chief export market. At the same time a government document reported that scientists had detected a marked decrease in frogs: the breeding stock was not returning to the shores of the lake.

The figures were immediately contested by the industry whose hired scientists argued that the census should have counted eggs and tadpoles not frogs. The tadpoles of today are the frogs of tomorrow, they said; our data show there will be a good run of frogs next year. Nevertheless, a total ban on collecting frogs was ordered by Ottawa.

At once the provincial government accused the federal minister of mishandling the frog stocks and hired a Newfoundlander to prove it. The commercial froggers pointed the finger of blame at the weekend sport-froggers, also expressing their belief that war should be declared on loons and diving ducks that undoubtedly were eating the tadpoles. The big operators said it was the fault of the little inefficient operators, adding that the government should buy them out and consolidate their licences. It's the Americans, said some; they're intercepting Canadian frogs returning to spawn. It's the Natives, said others; they're selling frogs commercially when they're only supposed to take them for religious and cultural purposes. It's not the fault of the technology, said the CEO of the Rana Reapers Manufacturing Company. It's like guns; reapers don't kill frogs, people kill frogs.

It's deadly ultraviolet radiation due to the thinning of the ozone layer, said the biologists: we're finding frogs with green sunburns everywhere.

It's clearly over-frogging combined with loss of marsh habitat, said the environmentalists: frogs are on the way out just like cod and salmon. We are going to take our anti-frogging campaign to Germany and put real pressure on the feds. Further, farming frogs is not a viable alternative, as they will certainly introduce diseases to the wild stock.

It's the environmentalists, said the froggers; they've got the government to set aside 12 per cent of the lakeside marshes as nature preserves, and that's where all the frogs have gone.

Soon a SHARE group was formed, arguing for wise use and equal access to every part of the lakeshore for frogging, and little signs appeared here and there in the village reading: "I Support the Frogging Industry" and "Frogging Feeds My Family." The latter appealed especially to the prolific family-values group of the Reform Party who bought TV time to argue that anything feeding a family must be good. Elect us, they said, and we will not only open the national parks for frogging, but also reduce taxes, pay off the national debt, increase military expenditures, and bring back capital punishment.

About this time the government ran another course for entrepreneurs and from it, as before, a brilliant idea emerged: Ecotourism! "Look," said the originator of the idea, a college graduate, "the frogging industry is a friggin' fiasco—its day is done! What we need is tourism and not just any kind of red-neck tourism where guys zoom around the lake in Seadoos and motorboats trailing their kids on rubber rafts. We want the baby boomers from the city, the wealthy guys, and what are they interested in? Nature, that's what, and the rarer it is the better they like it."

"The fact that we've exported everything means that we've got lots that's rare. We'll show them the last few specimens of the Slocan Three-Striped Frog in what remains of their natural habitat. Then we'll take them to a big Douglas fir or cedar or hemlock—I'm sure we can find a few back in the hills—and then we'll take them to the worked-out mines, and show them a bunch of old broken-down machinery. People love this historical stuff. They never ask whether or not it was all a big mistake. They'll go back to the city, happy, and leave big money behind!"

The idea worked for a few years. A few local purists complained that "eco-tourism" was an oxymoron because the tourists were mashing up the marshes trying to get pictures of the Slocan Three-Striped Frog. But just as "Save the Marshes" petitions began to circulate the tourism boom died down. Back in the city economic prospects were declining. With the mineral wealth and forest wealth of the hinterland shipped out, with plant and animal wealth exported and lost, with soils, water and air degraded, the city's life-blood dried up too. Computerized businesses dissolved. Stocks and bonds evaporated. Few resources remained to prop up city industry, nothing to justify the paperwork, the abstract facts, the figures, the bottom lines with which cities deal. Few now could afford expensive trips to the mountains for holidays. Times were tough for the leaky-condo set. In the capital city the Minister of Finance and his brain trust finally figured it out. "Export is not the problem," said he. "The problem is people, people scattered in small communities all over the land, a drain on the economy. Let us bring them all into the cities where they will have to be productive. We will rescue them from the idiocy of rural living."

And so by withdrawing funding for hospitals, schools and other social services out in the hinterland, the country people were forced to migrate to the crowded cities where they lived happily ever after.

ENVIRONMENTAL AWARDS THROUGH THE AGES

The decision of the village council to support a provincial Environmental Award for the team carving chunks out of the old-growth forest on New Denver Flats (reported in *The Valley Voice*, 18 March, 1999) puzzled me at first. But a bit of research into the history of Environmental Awards made it all come clear. The Award was for neatness, not a stick was left standing, and Council was simply following precedents.

Take for example Mary, Queen of Scots, who lost her head on the chopping block in 1587. At first the public was a bit nettled, muttering "That's no way to treat a lady!" But then the London Council called attention to the efficiency of the execution. One whack and Mary's head tumbled into the basket. It was a clean, clear cut, and soon disgust gave way to admiration. When a member of the Axemen's Union suggested an Environmental Award for the proficient Headsman, Council gave unanimous support, thus encouraging others to polish up their chopping skills.

Or how about a famous American case from Wild West days? At the conclusion of Custer's Last Stand, the Brave who removed the General's top-knot with a tomahawk was the recipient of an environmental award. When the news first broke many expressed the view that scalping was wrong. But Councillors invited to view the results found no ragged edges around the entire perimeter of Custer's cranium. In barber's lingo it was "a real neat cut," meriting approval and an Award.

Lastly, remember the woodsman on BC's West Coast who chain-sawed the famous golden spruce, a tree sacred to the Haida people? While some quibbled that a tree of such beauty and symbolic value should not have been destroyed, others stressed the superb job of felling that got the feller short-listed for an Environmental Award. It was pointed out that the stump was cut commendably low with very little "holding wood" so the five hundred and fifty rings could easily be counted. Simply put, a fine clear cut! Unfortunately the Haida Council has been unable to lure the man out of hiding to receive his just award.

I am pleased to share my research so that everyone can understand why the village council endorsed an environmental award for the local clear-cutting team, "notwithstanding the previous resolution opposing logging in Denver Flats." Council's indirect encouragement of clear-cutting forays into other domestic watersheds in the valley should also be noted. The main thing is to keep those clear-cuts neat.

From this a lesson for children who aspire to win awards, environmental or otherwise:

"It ain't what you do, it's how you do it!"

UNEXPECTED RELATIONS

Looking through the local paper, *The Valley Voice*, the other day I happened on an article about a new clinic opening in Nakusp called Planned Parenthood. I didn't have time to read it through, but what caught my eye was an end note that said, "If you've been drinking, and had unexpected relations, please come in and see us." I made a mental note.

New Denver's a beautiful mountain village on the shore of a crystal-clear lake. Those who live here know that as soon as friends and relations discover it, especially if they are from the dry prairies, they are apt to come back again, and again. "We were just passing through," they say, "on our way from Winnipeg to Vancouver, and thought we'd drop in for a day or two."

So it wasn't too much of a shock last weekend when I was drinking a beer in the back yard to suddenly realize that I had unexpected relations. "We were just passing through," they said, "on our way from Vancouver to Winnipeg and thought maybe you'd be home. . . ."

They looked hungry and there wasn't a thing to eat in the house. I hadn't vacuumed for weeks and the bed wasn't made. Then in a flash I remembered the newspaper item. I sat them down in deck chairs, excused myself, found the clinic number and dialed. A friendly voice came on the phone.

"Look," I said, "I'm fifty klicks away from you in New Denver but I've been drinking and had unexpected relations, so what'll I do?"

"Please come in and see us," she said. "Your privacy is absolutely confidential and you're welcome to bring your friend if it makes you feel comfortable."

"Well, there's three of them," I said. "Can I bring the whole gang?"

A long pause. "You mean you've had *three* relations?" She gave a merry laugh and a snort. "You can hardly call that 'unexpected!'"

I didn't like her levity. "What I wanted from you," I said in injured tones, "is advice on what to do about my relations."

"Well," she said, and I had the feeling that something was amiss, "we offer free condoms, but they're only effective before the relations. After that there's abortion . . ."

It was my turn to be funny. "Yeah, yeah," I said, "abortion is the answer, but only if it's made retroactive. Hey, that's the kind of recall we need for our local MLA!"

She cooled right off, obviously didn't get it.

"It sounds to me as though you've had one too many," she said primly, "we can't help people with your kind of problem," and she hung up.

I went back to my unexpected relations, we all had a beer, and I booked them into the Sweet Dreams B&B.

NATURE AND ME

"There is a special period, the little-understood, pre-pubital, halcyon, middle age of childhood . . . between the strivings of animal infancy and the storms of adolescence—when the natural world is experienced in some highly evocative way, producing in the child a sense of some profound continuity with natural processes." Edith Cobb.[1]

Sometimes in daydreams I am back in southwestern Alberta, in the little town of Granum where I grew up. There on the wide plains, rimmed on the west by the Rocky Mountains and their rolling foothills, the wonders of the world struck deep into me. Favourite memories are breezy summer days, slow drifting cumulus clouds, the meadowlark's song swelling and fading in the pulsing air, wild flowers and grasses nodding and waving in what we called "the prairie"—an unploughed strip of native grassland at the edge of town. There I'm running full tilt in pursuit of a winged jewel, ineffectively swishing a net, engaged with my butterfly partner in a zig-zag dance choreographed, now I know, by an inborn love for this Earth world.

The childhood exhilaration of immersion in Nature's wildness is well expressed in Wordsworth's "Lines Composed a Few Miles Above Tintern Abbey".[2] In memorable verse the poet conveys the aching joys and dizzy raptures inspired by the colours and the forms of Nature. But then, gently, he disparages such oceanic feelings as "the coarser pleasures of my boyish days." In maturity he "learned / To look on nature, not as in the hour / Of thoughtless youth, but hearing oftentimes / The still, sad music of humanity."

But here a question: Is sensitivity to humanity's still, sad music necessarily an advance over sensitivity to Nature's exuberant orchestrations? Was the aging poet perhaps expressing a convention expected of him by the society of his time, a society engrossed in getting and spending?

Those fortunate to have had happy childhoods know what losses are suffered in growing up. The emotional intensities of early pleasures fade, overlaid and submerged by schooling that disparages direct experience. Facts are taught, the abstractions of the "three Rs," and other necessities to prepare for "earning a living." Industrial society sets walls around us, mirror-like barriers that reflect back only the built environment and its artifacts, excluding whatever humanity itself has not fashioned. Species-centred culture turns vision inward to personal and social issues, distracting attention from our true lifelines: the ties to Earth instinctively sensed by the young.

I cannot believe that the blissful elation aroused in me by such magical places as "the prairie," and farther west the Porcupine Hills and mountainous Waterton Lakes, is fairly described as "the coarser pleasures of my boyish days," nor that the eclipse of these feelings by humanity's discordant sounds is the necessary mark of a maturer vision. I believe that childhood feelings for the Earth and its marvels should be nourished and accepted as fundamental truths in their own right. They should not be watered down with tears for the sad music of humanity. Concern for our own species is instinctive and will always flourish. That fact should make us prize whatever altruism, sympathy and love for things other than humanity we discover in ourselves.

My good fortune has been never to lose the feelings of connection that long ago I felt when looking on Nature "as in the hour of thoughtless youth." The sights and sounds, tastes and odours of the non-domesticated world, the sensual feel of it, have continued in adult life to spark an inner joy year after year. For I was fortunate in my employment as an ecologist, encouraged and financially supported to do what I love: roaming the wild, exposed to the magnificent landscapes of this continent for at least a part of every year, exploring its grasslands, forests, peatlands and tundra, inspired by a sense of wonder about it all. True, my feelings changed over time, though not in a Wordsworthian drift toward the species-focus known as "humanism." Rather, a deepening awareness of the profound significance of these other-than-human experiences grew in me, urging an Earth-centred faith and philosophy.

Grassland Experiences

It was Darwin's opinion that travellers ought to be botanists, because the flowering flora give character everywhere to the world's landscapes. This view was unknown to me when, happily, botany serendipitously chose me. Near the end of my freshman year at the University of Alberta, Dr. Ezra H. Moss, the only professor who expressed a personal interest, laid a friendly hand on my shoulder one afternoon in the botany laboratory. "Would you like to be a botanist?" he asked. I had always known I wanted to be a "naturalist," not understanding exactly what that meant except that it involved companionship with wild things out-of-doors. Dr. Moss's interests in the flora of Alberta and the ecology of its plants and plant communities indicated that botany could take me toward that goal. Then and there the decision was made to join and explore the green world.

Student travels followed, three marvellous summers in the grasslands of southern Saskatchewan and Alberta. This was a search in the closing years of the dry '30s for unploughed grasslands that could be joined with abandoned farmlands to make community pastures. I learned the names and habits of prairie herbs and grasses and much more: the beauties and mysteries of the prairie landscape with its sage-scented dawns, flaming red sunsets, crashing lightning-and-thunder storms that sweetened the air and faded away as quickly as they had come. High on windy hills we found old teepee rings sinking into the turf, each boulder encircled with white-flowered moss phlox. On the hardpan of eroded fields, among sparse yellow sweet clover, stone-age spear points and arrowheads of chert and obsidian. In lonely coulees and draws where silvery atriplex indicated a salty soil, bacculites and giant ammonites eroded out of ancient sea-bottom shales. In such rovings, rare horned toads, rattlesnakes, prairie dogs and sage grouse crossed our paths. Songbirds of the south, formerly unknown to me—bobolinks, lark sparrows and lark buntings—flashed their black-and-white patterns in fluttering flight and sang to me from wire and fencepost.

Occasionally I came across tracts of native grassland so large that, hiking within them, all jarring reminders of civilization faded away. Far from the towns and cities, far from the truck and tractor, feelings of a peaceful, timeless quality gradually suffused my consciousness with vague longings, tuggings at the heart. Gazing around, questions arose. For how many millenia had this part of Earth been just so: no fences, no poles, no buildings, no herefords, nothing but waving grasses from horizon to horizon, year after year after year. What feelings of long, long time. Here was history's real stage; here was the world "made flesh." The historic trivia of kings and priests and wars washed out of the mind, replaced by imaginings of all that had transpired *in this very place, on this piece of Earth*. Puerile human deeds of derring-do faded to their rightful insignificance.

Such large unplowed tracts were rangelands, used at certain times of the year for grazing cattle. The belief was strong, then as now, that any piece of Earth's surface not producing something of human worth was barren, indeed was *wasteland* or, in the insensitive language of agriculture, *unimproved land*.

Once as I walked through such a rangeland, carrying only a metre-square frame for estimating grass abundance, a herd of coal black horses suddenly materialized, streaming up out of a coulee half a mile away. Immediately they saw me, a strange two-legged object in their four-legged world, and with manes flying they galloped straight at me. No defence, no escape, no place to go. Just before I was certain to be trampled, they drove their hooves into the turf and stopped, forming a half circle around me, neighing, tossing their heads, whisking their tails. Fearful but thrilled, my mind made a flip and I was back on early Pleistocene plains confronting *Equus* ancestors, sharing with them a wild and exciting moment. Just over the rise, in my imagination, a saber-toothed cat lurked, perhaps a pack of lobo-wolves, camels, giant ground sloths, elephants—all the ancient species that once roamed here. . . .

The momentary illusion faded. The child of African savannas waved his measuring frame at the children of North American grasslands.

Mammalian relatives, both species born of open landscapes, we eyed each other cautiously and then, curiosity satisfied, the horses snorted and ambled off to graze. The danger past, I looked up into the blue sky where the native skylark, Sprague's pipit, sang his sweet descending cadences, gazed around at the low hills with their varied shades of green dappled with flowers, and knew with healthful certainty that nothing in life could be better than this.

On a similar occasion on a high hill I remember facing into the western breeze near dusk as long shadows crept over the landscape hollows and the setting sun painted stray clouds gold and old rose, moved to sing something of my lonely gladness. The expression of mood that came into my mind was, "You are the promised kiss of springtime that makes the lonely winter seem long; you are the breathless hush of evening that trembles on the brink of a lovely song . . . the dearest things I know, are what you are." The ballad, lyrics-and-music by Jerome Kern, is as appropriate for the landscapes of home as for the dear ones we love.

Later I wondered why that particular song rather than a traditional hymn had come to mind. From childhood on I was familiar with the songs in the United Church hymnary. The inadequacy of hymns relates partly to the atmosphere of the places where traditionally they are sung—spare enclosures that purposely wall Nature out—and partly to their otherworldly focus. The old Greek and Jewish philosophers who inspired the hymns were unhappy with the impermanence and precariousness of this life. They yearned for an unchanging realm of pure spirit: a safe and certain harbour for the good soul after the storms of life. Thus the favoured hymns sing praises for an insubstantial hope. The exceptions, praising the beauty of the Earth, are mostly consigned to the section of the hymnary marked "For Little Children," prompted by the thought that the sensuous delights of the natural world (Wordsworth's "coarser pleasures of my boyhood days") are, like cuddly animal toys, permissible and harmless in childhood. For adults, sterner more abstract stuff is prescribed.

Post-graduate education in ecology changed my views and valuations of the species *Homo sapiens* and the world it seeks to dominate. My grassland studies began at the University of Nebraska in the city of Lincoln, central to the fertile prairies that extends northward from Oklahoma to Manitoba's Red River Valley. The scholarly ecologist Frederick E. Clements named the region "True Prairie," for the very good reason that he began his studies there. His first love was naturally "the true," the norm, the standard for comparison with the moist Tall-grass

Prairie to the east (now cornland) and the drier western Mixed Prairie (now wheatland). Despite its intermediate position, its "golden mean" status did not save the True Prairie from agricultural conversion, and by the 1940s its unceasing destruction was torturing the heart of my mentor, Professor John J. Weaver.

"It has gone extinct" is a favoured way of referring to a thing we have destroyed, implying no human assistance and hence no blame. As thesis study-area, Weaver assigned me to one of the last remaining prairie fragments just before it "went extinct." Though close to the city of Lincoln, it had miraculously survived surrounded by ploughed fields: a spectacular half-section of head-high grasses and herbs sheltering a myriad of lower plants, birds, and small mammals. All through my spring-to-autumn study it flowered in colour co-ordinated waves, beginning with the delicate whites and blues of spring ephemerals and ending with the strong purples and yellows of asters and goldenrods. Shortly after I completed my thesis the prairie disappeared, replaced by a muddy black field slated for crops of corn. The positive side of the loss was another small prompting to care for the lovely things of Earth that have no cash-in-hand utility.

Ecological Understanding of Earth as the Metaphor for "Life"

Chance reading in the library opened my eyes to the philosophy of Alfred North Whitehead who extended the idea of "organism" to all reality. If organelle, cell, organ and organism are recognized as successively higher orders, why not carry on to conceive the next upper level surrounding organisms as also an organized and integrated thing: as an *ecosystem*? If every organ exceeds its cellular parts in complexity, and each organism is more "organized" than its organs, should not the sectors of Earth of which organisms are parts transcend them in intricacy, in creative potential, in importance? And ought not *ecology*—the outward-looking science—to attend to these larger entities in which we too are embedded?

Ecology is the science of context, a persistent reminder of the importance of what stands outside every object of interest. Ecology teaches the absolute dependence of things on what is peripheral to them, a necessary counterbalance to popular kinds of reductionism—scientific, psychological and sociological—that pull attention downward (toward atomic explanations) and inward (toward humanistic explanations). Earth's sectoral parts—air above, water or land below—constitute each organism's context and the matrix in which we humans exist. From such ecosystems are organisms born: horses out of American grasslands, people out of African savannas. We are Earthlings, Earth internalized, humans from humus. Without the surrounding world of Nature our bodies could not be, nor our minds, emotions, languages, and cultures.

The very idea of life is changed when placed in this perspective. Are we alive because our organs and cells are keeping us so, or is "aliveness" a gift from Earth that maintains the functions of these less complex levels of being? We know that vitality quickly fades when organisms are deprived of their appropriate environments. Can we not attribute the life-force at least as much to what surrounds us, as to some unknown internal essence? Life is in and of the Earth, and each organism, organ and cell partakes of it as long as energetic ecological lifelines are open and functioning. We protect life by protecting Earth and its ecosystems and, with the same protection, biodiversity is preserved.

Mountain Forests

Half a century after my true prairie sojourn I live in a valley in the interior of British Columbia. Ideas sparked years ago by grassland experiences are equally relevant in this moister milieu of rainy clouds draped over mountains, of lush forests reflected in a crystalline lake. Each year from April to June the thickly wooded slopes replenish the lake, feeding it filtered rain and snowmelt through hundreds of small streams that provide good water for wild animals, as well as for the self-domesticated one. Here as elsewhere in the world, humans deceive themselves about their own importance and the relative unimportance of things exterior to them. Just as on the prairies, and in the trans-Canadian boreal forest, the natural landscape ecosystems of British Columbia are under attack.

The motive—a cultural norm long taken for granted—is to increase the human share of Earth's wealth by taking from land and water everything of worth, while discarding whatever does not directly serve us. In the Slocan Valley this means replacing diversity with simplicity, substituting blocks of even-aged plantations for the uneven-aged natural forests, cutting out the slow-growing conifers and replacing them with such sprinters as pine, Douglas fir, and western larch, converting entire landscapes to what are essentially tree-farms. Clear-cut logging is the approved technique both because it is "efficient," and because it industrializes the landscape as a future wood factory. Clear-cutting in large tracts is perfectly suited to big labour-replacing machines, making it least costly. Beyond its shoddy cheapness, clear-cutting prepares the ground for monocultural planting of genetically selected trees, a necessary step in the design of industrial forests whose single goal is provision of a steady supply of fiber for the mills. Along with its road-building and trucking, this violent treatment of forested terrain disfigures and destroys the beauty and integrity of mountainsides, quickens the erosion of soils, reduces biodiversity, and fouls surface water. These externalities are judged relatively worthless.

The valley's good water provides no jobs nor corporate dividends, and so it gets little protection. What gets protection, to the detriment of all else, is the annual allowable cut that keeps the mills running and supports a scattering of jobs. Concerns for the maintenance of forest ecosystems in all their complexity, for the preservation of aesthetics, wildlife habitat and clear water, are not factored into the profit equation. What remains of the primeval forests is sacrificed for the stock-holder and the pay cheque. As Don Kerr has quipped, "Beauty is in the eye of the shareholder." The academic world says little by way of protest, or justifies the mayhem with the postmodern thesis: language makes reality; all is discourse, and therefore Nature is whatever the majority decides it wants. When the forests are gone, Disneyland—the choice of the urban majority—will adequately serve humanity's needs.

A Pilgrimage

In late October with the forest colours changing, I sat down beside the timeless lake to admire the sunlit sheen of the water and, rising out of it on the far west side, the Valhalla Mountains. Near the shoreline, reflected in the water, groves of yellow-leaved birches form an undulating band, rising on the lower mountain slopes where soils are deep and moist, falling toward the lake or disappearing altogether where the soils are shallow. Displayed between the birchy patches, extending upward on the rocky mid-slopes, are the mottled greens of hemlock, western red cedar, Douglas fir and western white pine. In the old mining days these forests were selectively logged using horse-power, and now with protection they are allowed to be wild again, which makes me glad. Higher on the slopes the green matrix is dappled with dull orange dots—an impressionistic "pointillism" of western larch crowns changing their colour before leaf fall, and a prophetic warning of fires to come. Then the slope abruptly steepens, cliff faces appear and with them a different darker forest that, in profile on the skyline, shows the unmistakable narrow spires of alpine fir. Above this subalpine forest, at the highest mountain levels, craggy peaks enclose an ice-carved cirque in which nestles, pure white, a remnant of the last great ice age: the New Denver glacier. Climatic warming is reducing it to a summer snow-field.

I have been to the foot of the glacier several times, once in 1945 and twice again in the last few years. The first time up, our little company of three met a grizzly bear coming down the path, an experience that startled the bear as much as it startled us. Later, off the trail-side and well below us in a little green meadow, we saw a female romping with her two cubs. At our high-altitude campsite, just below the glacier, a clan of mountain goats came unafraid and inquisitive, wondering who these strangers were, checking us out much like the black horses in Saskatchewan. Hiking the same terrain half a century later, no signs of bears nor goats were seen although they have not "gone extinct" quite yet.

Recent trips were pilgrimages, necessary to clear my head, to breath again the invigorating fir-scented air, to look down into our beautiful valley from the subalpine and alpine zones, to continue the dance with Nature that has been my joy and recreation each year since those early Granum days.

My last climb was in late September, Indian Summer weather—quiet, hazy and warm. Gardening was nearly over for the year and just before departure I picked the last errant spike of lupine, a late-bloomer that I had seeded in the spring, placing it in a jar of water on the kitchen table. Although I had examined its purple blossoms carefully for hitch-hiking fauna and was satisfied that nothing but lupine had been collected, soon there appeared on my table-cover two inch-long slugs who straightaway oozed off in search of a better land. Pondering on the wit of these restless wanderers who relinquished velvety, sweet-scented flowers for a flat plastic landscape, I headed for the lakeshore and loaded my canoe.

In the calm of the morning the lake was a mirror and I a sacrilegious scratcher of it on my way to the little delta at the mouth of Sharp Creek—the discharge point of meltwater from the glacier some eleven kilometres beyond and above it. As soon as I hoisted my pack and started up the creekside trail the old excitement came back, a sharpening of the senses of sight, scent and sound. The resilient needle-covered path, quieting my footsteps, meandered through a mossy old hemlock forest with a scattering of western red cedar. Then a steep climb up into the first hanging valley shaped like a ladle, its "handle" a roaring waterfall on the bedrock backwall. Where the trail leveled off I rested and watched a pygmy shrew darting back and forth on sticks that had fallen into the stream, sometimes on top of these broken bridges, sometimes underneath, persistent but unsuccessful in locating a dry path across. The metabolic rate of this tiny carnivore is said to be so high that it must eat every four hours or starve to death, a thesis I had not the time to test. When last seen, its frantic pace betokened a determination to find the missing passage or, like macho Arctic explorers, to perish in the attempt.

On again through a mixed forest of Douglas fir, western larch, lodgepole pine, western white pine, white birch, and aspen, with a sprinkling of old charred snags, for these are trees that thrive after lightning fires. Then across the toe of a slope composed of huge stone blocks, prized from the cliff above by frost action, and the first muted alarm cries of a family of pikas, the second cutest animal in the world (first place is generally accorded to pandas). These rotund little "rock rabbits," sometimes called "conies," have rounded ears, big eyes, and fur-soled feet for traction in scampering over the rocky talus that is their usual home. Their calls remind me of the vocalization of nuthatches, the same nasal "yank, yank, yank" with something of a ventriloquist quality, difficult to imitate. During the short summer they gather sedges, grasses, and other flowering herbs for winter provender, spreading them out to be cured in the sun: hay-on-the-rocks for food. What, I wondered, do they drink on-the-rocks?

Onward up a switchback trail to the head of the waterfall and a cautious look over the edge into the foaming depths. Here where the air is cooled by spray from the icy stream, a few Engelmann spruce and alpine fir have found an environment to their liking below their usual range. Hoist the pack and onward through another "fire forest"—western larch, birch, poplars, Douglas fir—and finally, over the lip of the second bowl-shaped valley, easier walking again. Camp-site for the first night was a streamside grove of giant western red cedars, hundreds of years old, at too high an altitude to be reached by the early horse-loggers. Lying in my tent at night it occurred to me that camping beside a clamourous mountain stream is far from ideal, as all other noises are obliterated: no wind sighing in the trees, no owl hoots, no grizzly bear grunts—only endless white-water babble.

On the second day I left most of the weight behind in the tent and went on with a light pack. Now the montane forest is left below—no more Douglas fir, hemlock or cedar—as the climber ascends the side wall of this second hanging valley which, like the first, has a waterfall on its nearly vertical back wall. The path zig-zags up a broad avalanche slope covered with pliant shrubs—alder, mountain ash and willow—able to withstand the rush of each winter's avalanches. Lush patches of cow parsnip or Indian celery, prime food for grizzly bears, warned me to advance slowly and keep a sharp lookout.

Once more into wooded terrain, but what a difference at an altitude of about 2000 metres! Here the air is cooler, sharper, filled with the fragrance of evergreen alpine fir and Engelmann spruce with their heathy understory of false azalea and white-flowered rhododendron. This fine sub alpine fir/spruce zone is the mountain equivalent of the great boreal forest belt that sweeps across northern North America, from Alaska to Newfoundland. One walks through its summer-scented glades as if enjoying a stroll through the highlands of the boreal Gaspe in La Belle Province.

Nearing the mountain top, after picking my way over several recessional moraines of jumbled giant boulders, I came into the upper cirque and to the base of the glacier where Sharp Creek bubbles out of the rocks under an awesome ice field. To my surprise the glacier was channelled with huge crevasses, running downslope, dividing the ice mass into blocks. In the face of climatic change, this small glacier is on its way to extinction. How changed in size since my first visit in the early 1940s! Below the glacier the cirque floor has been flattened with glacial "flour"—silty soil filling in what must have once been a little tarn. Now it is covered with shallow peat, topped with a hummocky sedge meadow through which the new melt-water creek meanders. At its rocky outlet, before the first little waterfall, a plump grey dipper bobbed and sang, and watching through binoculars I saw it wink white eyelids at me. But now the sun was lowering, a warning to begin the descent.

Just before entering the forest on the downward path I heard shrill whistles that at first I took to be human. Someone trying to get my attention? Yes, the questioning signals, from about fifty metres below me, came from three stout marmots: "whistlers" or "siffleurs" in the more appropriate argot of the Francophone. The biggest one, lying flat out on a boulder, resembled a sunning grey seal. We exchanged whistles for fifteen minutes while I admired them and they wondered whether I was hungry or just passing by. Then down again to base camp in the cedar grove, and to the lakeshore on the following day where a wind-storm and high waves detained me overnight.

The Wild Needs Protection

Early next morning in the pre-dawn calm I launched my canoe toward the settled shore, turned my back on the wilderness, dropped away from its alpine meadows, subalpine and montane forests, waterfalls, avalanche tracks, the shrews, pikas and marmots, and made my way back to the plastic world. Unlike the slugs in my kitchen who deserted living blossoms for a synthetic table-cover, I have an inkling of what we give up when we forsake the wild world, a feeling for the losses suffered when we ignore it as of little worth and, based on the same mistaken valuation, unthinkingly set about annihilating it. In a deteriorating world, I want all the wild places that are not yet severely harmed to be there next year, and for years after that, not gone forever like the various vanished "prairies" I have known, a wealth of beauty spots whose memory will vanish with me.

The world, the Earth, is good and supremely valuable. We err when we suppose that we are God's special creation and therefore worth more. What must be opposed is the pernicious belief that the universe is human-centred, that all else on Earth in land, sky and water is of lesser value than human life. No divine providence has given us the right to plough, mine, slash and burn, displacing and exterminating all organisms except our own kind, tormenting the paradise into which we are born, often only to satisfy frivolous wants.

I dream of a culture that will let wild ecosystems be, free to do their own creative things, to make their quiet contributions to Earth whether or not this unimportant I-ego and others like it are around to chase butterflies, admire prairie flowers, salute wild horses, climb mountains, sing with pikas, whistle with marmots, and otherwise applaud the great "Not-Ourselves."

References

[1] Cobb, Edith, "The Ecology of Imagination in Childhood." *The Subversive Science* eds. Paul Shepard and Daniel McKinley, (Boston: Houghton Mifflin Company, 1969) 123-124.

[2] Wordsworth, William, "Lines Composed a Few Miles Above Tintern Abbey" *The Norton Anthology of Poetry* ed. Arthur M. Eastman (New York: W.W. Norton & Company, Inc., 1970) 241-244.

2. FOR THE BEAUTY OF THE EARTH 

FOR THE BEAUTY OF THE EARTH

A Talk to the Vernon Natural History Society

What a pleasure to come again to the splendid Okanagan Valley, especially when at its spring best! My first visit more than half a century ago will never leave my mind because, as I hitch-hiked south from Vernon, I saw my first Western Rattlesnake. Head up and tail rattling, it was signaling "beware" to a circling dog. Farther along, by the roadside, that wonderful flower the Scarlet Gilia, or Skyrocket, swayed with the grasses under impressive cinnamon-barked Ponderosa Pine. And pink-bellied Lewis's Woodpeckers, to my astonishment behaving like flycatchers, coasted out from the pines to pick insects from the air. The clear green colours of the lake were marvellous, and new landforms appeared: creamy silt terraces on the valley walls marking the levels of ancient glacial lakes, and rising behind them savanna-clad rolling hills. How fortunate you are to live in a world of such marvels, not least of which (so I've been told) is a great blue heronry in a tree behind fast-food McDonalds! Who says we can't have the best of both worlds: hamburgers and wildlife?

On today's field trip it was obvious that everyone is intrigued by birds. Their ancestral stock and our mammalian stock branched off from the reptile-dinosaur line in Jurassic times, 190 to 140 million years ago. Somewhere in the same period this piece of Okanagan terrane on which we stand, originally a large off-shore island in the Pacific, collided with the westward moving continental plate, adding another piece of real estate to British Columbia and crumpling up the Selkirk Mountains to the east. Doubtless we are all related—crustal plates, landscapes, the feathered and the featherless bipeds. Made from the Earth, we people share with the birds genetic DNA, a diaphragm for breathing, a four-chambered heart and two legs. Also our brain-stem, the medulla oblongata, is the original "bird brain"—a term sometimes inappropriately applied to our neighbours. Remember that Minerva, the goddess of crafts and wisdom, took the wise old owl as her mascot.

The comeliness of living organisms, plants and animals, draws our attention and interest. Landscapes too with their many intriguing surface forms, overhung by the cloud-decorated sky, contribute to our sense of the beauty of the Earth. The instinctive attraction to Nature's elegance leads to membership in natural history societies and various other kinds of out-of-door clubs in every nation. When living on the prairies I belonged to an elderly Nature-loving group called "The Golden Eagles" (whose female members mischievously suggested "Bald Eagles" as a more appropriate label for those of the alternative sex). Just as on our foray today, we Saskatoonians went out exploring—intrigued by the different landscapes of hills, valleys and plains, admiring flowers in season, watching for wild creatures. It was fine recreation, good exercise in the fresh air, with a relaxing lunch together in a quiet spot. Few thought of it as anything more than an enjoyable outing, but I am sure that it goes deeper than that. In such experiences we return to our evolutionary roots, exposing ourselves briefly to the way we were, to the way humans lived as foragers for most of the last four or five million years, ever since the genus *Homo* branched off from the chimpanzee and bonobo apes.

Field-tripping—wandering in the flowery hills, watching the birds—is our evolutionary history. Like the other primates, our ancestors for hundreds of thousands of years were walking-trotting foragers—gatherers and hunters. Our two best doctors are our legs. The settled life in communities, on whose perimeters agricultural crops are grown, is a recent way of living perhaps no more than nine thousand years old. And until about one hundred years ago, most of our forebears lived close to the land, in the country, not the city.

Some argue that the evolutionary history of the human race, how our ancestors lived for hundreds of thousands of years, no longer matters. They say that where we came from, what we used to do as foragers or more recently as simple agriculturists, has no bearing on what we are today and what we will be tomorrow. According to this theory we are fundamentally creatures of culture, making ourselves in whatever pattern suits the modern mind. If people continue to flock into cities, if they live most of their lives in the cocoons of buildings and cars, if they isolate themselves and their families more and more from others, bunkered in with a computer and TV set, if they accept the virtual reality of videos and movies as just as good as or better than the reality of forests and rivers, mountains and sky, then that's OK. Few are complaining, so all's well. Where's the problem?

I don't buy it. Certainly culture is important, but culture—the taken-for-granted system of beliefs and activities by which a society lives—is relatively recent. It is learned and changeable, therefore superficial compared to our shared biological nature. An analogy is the skin that keeps us from drying out and contributes to our good looks but hides from view the essentials: the breathing lungs, the beating heart, the digesting stomach, the image-making brain, and all the other organs and organ systems including the skeleton. Culture, carried and transmitted by language, can also conceal fundamentals. When narrowly concerned with the human species as individuals and societies, ignoring their Earth-born status and dependencies, culture hides reality from us.

We know from history that many cultures before ours were not sustainable. The downfall of civilizations not weakened by wars usually resulted from neglect of their environments, which leads to the suspicion that we too are on a disastrous course. In the interests of efficiency, speed, comfort and provision of consumer goods far beyond our needs, Western culture entices us away from our biological-ecological roots. What the media call "environmental problems" are really people problems: the overuse of Nature's wealth in forests, agricultural soils, the oceans. Increasing pollution and decreasing biological diversity are signals that many of the fundamental cultural beliefs guiding our actions are wrong. Were our values right and appropriate for the places where we live, were we living in tune with the Earth, then "environmental problems" would disappear and general human welfare would improve.

Western philosophy, religion, science and technology have developed in ways that separate us more and more from the natural, non-humanized, non-domesticated world. Not so long ago we were natives, at home in the Ecosphere, but now many people feel like strangers on Earth, fearful and alienated. Ecological psychologists diagnose the general malady as EDD, Earth Deficiency Disease. Too much culture, not enough Nature. Where is the cure? Consciously choosing to reconnect with Earth's great out-doors seems to me to be a good start. The more we understand and appreciate Nature, the less our fears. With practice we may find our way back to the lost place where we have always been.

A few years ago someone told me that in Spain the school curriculum was revised so children could be taught not only human cultural history but also natural history. If true this is revolutionary. Imagine, human species studies and Earth studies fifty-fifty! Not just an emphasis on the achievements of *Homo sapiens* but, equally, an emphasis on the ecological history of humans as Earthlings, on the source of their life in Earth, on the dependencies of all organisms on others as well as on the matrix elements—land, air, water.

I daresay that the natural history of the Okanagan Valley is a very small part of the school curriculum here from Kindergarten to Grade 12. Probably few children know the wonderful story of this particular ecoregion, how it was formed, what plants and animals are adapted to live here, what they do in their life cycles, how the native people lived as a part of it all, and how dependent on it the residents of the valley still are today. Some may defend this gap in education, asking: "If natural history were given a more important role in education, would it help graduate students into the job market?" The honest answer is "No." On the plus side, a better acquaintance with natural history, fostering an affection for the landscapes and waterscapes around us, could be important to the survival of thousands of species including, over the long run, our own. Contrary to popular opinion, economic success in the job-market is not the only, nor even the first, measure of a successful education system.

In his book *Biophilia* (love of life) the scientist-biologist Edward O. Wilson suggested that what we do in the name of recreation—nature hikes, berry picking, fishing, bird watching, gardening, keeping pets and house plants—reveals fundamental human needs. We are deeply connected to Earth and its life-forms, and feel energized in natural settings. A common emotional bond is imprinted in our genes, he says, making us affectionately responsive to the landscapes of home. We need to acknowledge and respond to these deep feelings, nurturing a relationship to Earth based on respect, humility, cooperation, and love.

It used to be that the followers of Darwin and Einstein allowed themselves only the statement of facts. They shied away from expressing value judgments. Now, more and more, the best ones are exhorting us to show our feelings, get emotional about important issues, express love for the attractive things of Earth. It's good to see the change. One scientist, having noted in the journal *Science* that humans "have expropriated 95 per cent of Earth's surface from other species," closed his article with this statement:

> The land remains. Share it more generously with other species. Do the research to discover gentler ways to occupy the land, ways to reconcile our uses with those of the many species that also need it to sustain life.

If all people, children and adults, were more exposed to the planet's beauties, if we understood Earth better and were taught to appreciate and love its creativity, its co-evolved creatures, its cycles of birth-death-rebirth as the seasons come and go, we could not help but change our ways. We would be inspired to tread more lightly on Earth, to shrink our "ecological footprint" by curbing growth, to break the fossil-fuel addiction that is the chief source of pollution and climate change, to better manage precious water and our wastes, to conserve and restore damaged landscape and waterscape ecosystems. Putting Nature first would be more than second nature.

Separation is one definition of sin. The wages of sin, for separating ourselves as a species from Earth and its other creatures, are paid today by catches in our eyes, ears, noses and throats. A step toward mending the separation is to acknowledge and rejoice in the fact that we are Earthlings made from stardust, humans from humus, our bones from the old coral reefs uplifted in limestone mountains, our flesh from minerals of Earth's crust mixed with carbon and oxygen from the air. This is spirituality of a worldly rather than an other-worldly kind.

No need to feel guilty if you skip church for a walk in the woods. Your experiences of connectedness with the natural world are just as genuine, just as conducive to spiritual well-being as those experienced in the pew. This truth was emphasized by the Chinese sage and artist, Tsung Ping, who wrote in the fourth century AD:

> Landscapes have a material existence, and yet they reach into a spiritual domain. The wild beauty of their form, the peaks and precipices rising sheer and high, the cloudy forest lying dense and vast, have brought to the wise and virtuous recluses of the past an unending pleasure, a joy which is of the soul and the soul alone. One approach to The Tao (The Way) is by inward concentration alone. Another, almost the same, is through the beauty of mountains and water.

It is the outer vision, the comprehension of ecological spirituality, that today needs public attention. Inner spirituality is over-emphasized in our culture, often leading to an uncritical rejection of Earthly being—as if our actual body/minds were no more than shadows of a greater reality unperceived. Philosophers and theologians have weighted our thinking-feeling intelligence one-sidedly with abstract ideas that have no counterpart on Earth. Today we need open-eyed gurus, helping humanity to reconnect with Nature, restoring the spiritual balance.

Exposure to existence outdoors, outside society's technologic cocoons, is the way to foster emotional connections. Then add the knowledge that reinforces feelings. Of the two, sensory experience comes first. When we go out as today in May to admire the Earth—its rivers, lakes, plains, mountains, forests, and the new springtime life—there is a joyousness we feel in the air. Spring is the time of rebirth and renewal, impressing on us the mystery of growing and greening as the exuberant beauty of Earth is revealed again. All visions of paradise, on which ideas of heaven are based, draw for their descriptions on abundant and delightful flowers, fruit trees, running water, bird song—the Garden of Eden! Paradise is a beautiful garden.

In the early 1970s the first federal Department of the Environment was established in Ottawa, and with it an Environmental Advisory Council to give the Minister sage though unwanted advice. Wondering where to begin, the Council members decided to frame an Environmental Ethic that might give direction to the government and general public. What would be the strong peg on which to hang an Environmental Ethic? they asked themselves. What would rouse a sense of the importance of the surrounding world? What would be the grabber to catch the attention of everyone, regardless of ethnic origin, culture, language, gender? After long thought the Council decided that beauty must be the central theme. And so the Environmental Ethic of 1975 reads: *Every person shall strive to protect and enhance the beautiful everywhere his or her impact is felt, and to maintain or increase the functional diversity of the environment in general.* The jargon of the last clause about "functional diversity" was added as a sop to the Council's prosaic scientists.

Beauty, yes! We are all lovers of beauty, and beauty makes us lovers. Experience of the lovely and the sublime bonds us emotionally, affectionately, to Nature. In the words of the Ethic: We are moved to protect and enhance the beautiful everywhere our impact is felt. Those who believe there is no beauty as such, that beauty is entirely subjective "in the eye of the beholder," have not thought deeply about the universality of things all people admire nor, on the side of ugliness, what everyone decries.

Well ahead of the Environmental Advisory Council, Albert Einstein opined: "Our task must be to free ourselves from this prison by widening our circle of compassion to embrace all living creatures and the whole of Nature in its beauty." And a more contemporary voice, that of Stephen J. Gould: "We cannot win this battle to save species and environments without forging an emotional bond between ourselves and Nature as well—for we will not fight to save what we do not love." The Union of Concerned Scientists issued this statement: "What is regarded as sacred is more likely to be treated with care and respect. Our planetary home should be so regarded." The sacred is a name for whatever moves us deeply, emotionally.

The second way we make connections, through our thinking minds, is also important in backing up sentiments of love and respect for the Earth and its landscape ecosystems. There's the fact of knowing—thanks to the sciences of geology, biology, evolution and ecology—that we are Earthlings, created from and maintained all our lives by the surrounding reality. A helpful ecological fact: we could not, in any imaginable sense, exist without the Earth. Slippery language has equated "life" with things like us that grow, reproduce, and evolve, not considering that such miraculous processes are impossible without the surrounding matrix of air-soil-water. In the big sense, Life is Earth and Earth is Life. No Earth, no Life. No Earth, no people. Thus the higher value is transferred from humanity to the blue planet.

The conclusion I draw is that Earth's regions—those within which we pass our lives—should as much as possible be protected from ecoterrorist activities now proved destructive, such as clear-cut logging on mountain watersheds. The preservation of remaining wild and semi-wild areas is increasingly important, and everyone can contribute even in small ways—perhaps protecting a pond, a grove of trees, or a patch of native grassland on one's property. For those who are able, the bigger task is working for protection of large wildernesses, new national parks, networks of ecological reserves. Note that the emphasis is on preserving land and water areas, on landscape and waterscape ecosystems—the homes of all wild creatures.

When landscape diversity is protected then biodiversity, the diversity of wild plants and animals, will look after itself. No wild creatures can be preserved without first attention to their life-sustaining environments. Ideally large representative core areas of land should be protected, rather than tiny islands of green here and there. Nor should the concept "representative area" be confined to only one or two of every kind of terrain. Single landscapes, even double landscapes like animals two by two, are unlikely to survive. It may have worked for Noah, but he had powerful assistance lacking today.

The cost of living lightly with the Earth is a bargain, requiring only a reduction of our wants. On this theme Wendell Berry speaks wise words. Here is one of his poems diagnosing our society's illness, prescribing the cure, and ending with the words "practice resurrection"—which I take to mean reviving the damaged and degraded, restoring as much as we can the beauty of the Earth.

Manifesto: The Mad Farmer Liberation Front

Love the quick profit, the annual raise,
vacation with pay. Want more
of everything ready made. Be afraid
to know your neighbours and to die.
And you will have a window in your head.
Not even your future will be a mystery
any more. Your mind will be punched in a card
and shut away in a little drawer.
When they want you to buy something
they will call you. When they want you
to die for profit they will let you know.

So, friends, every day do something
that won't compute. Love the Lord.
Love the world. Work for nothing.
Take all that you have and be poor.
Love someone who does not deserve it.
Denounce the government and embrace
the flag. Hope to live in that free
republic for which it stands.
Give your approval to all you cannot
understand. Praise ignorance, for what man
has not encountered he has not destroyed.
Ask the questions that have no answers.
Invest in the millennium. Plant sequoias.
Say that your main crop is the forest
that you did not plant,
that you will not live to harvest.
Say that the leaves are harvested
when they have rotted into the mould.
Call that profit. Prophesy such returns.
Put your faith in the two inches of humus
that will build under the trees
every thousand years.

Expect the end of the world. Laugh.
Laughter is immeasurable. Be joyful
though you have considered all the facts.
So long as women do not go cheap
for power, please women more than men.
As soon as the generals and the politicos
can predict the motions of your mind,
lose it. Leave it as a sign
to mark the false trail, the way
you didn't go. Be like the fox
who makes more tracks than necessary,
some in the wrong direction.
Practice resurrection.

IN SEARCH OF THE HOLY GRASS

In the Native village of Fond-du-Lac, beside Lake Athabasca in northern Saskatchewan, a small group of us waited for the waves to subside before heading off to the magnificent south-shore dunes. The month was June, the wind was sharp but spring-like, and returning sparrows sang in the newly leafed-out bushes.

Around us, curious children gathered as they typically do wherever innocence has not been corrupted by adult and urban fears, and one of them eventually led me to a young man who spoke feelingly about his cultural roots. He disappeared into his small house and emerged with a braid of shining grass stalks.

"It's ceremonial sweet grass," he said, touching it reverently, "I got it from the Blackfeet in Montana."

We admired the fragrant grass, and presently I asked him to come with me to the lake shore to see an interesting discovery made a few hours earlier. We walked to a patch of sandy beach where a miniature forest of flowering grass stems shimmered and swayed in the sunny breeze. He looked doubtfully from the braid in his hand to the turf at his feet. Could this really be sweet grass, Indian grass, *Herbe Sainte*, the Holy Grass, growing practically on his doorstep? Sacred ground, in Fond-du-Lac of all places?

Most of us are dubious about finding the metaphorical Holy Grass near where we live. Perhaps it is because sedentary life in towns is not natural. There, sensory deprivation conditions us to ennui. We were meant to roam, to wander, to migrate with the seasons, and are soon bored with the familiar miracles on our doorsteps. We belittle them as "mundane," defined as "of this world"—a synonym for the dull and uninteresting that by rights should mean the wonderful, the marvellous!

Exiles from the Natural World

Even when we flee the concrete-and-glass jungle—taking to the countryside, the park, the woods, the lake or the mountains, hoping to find places of re-creation—the Holy Grass still evades us. The longing to find, to rediscover and enter an enthralling world of mystery and charm, is thwarted by taught ways of thinking that somehow tarnish all deep feelings, all moving affective experiences out in the wild.

When literate people such as we chance upon Holy Grass, our cultural perceptions are likely to mar its perfume and dim its shine. It is "just grass" whose attractive characteristics make it a likely candidate for the museum, displayed on a herbarium sheet and labeled *Hierochloe odorata* (L.) Beauv., a member of the *Phalarideae* tribe in the family *Gramineae*, named from the Greek *hieros* meaning sacred and *chloe* meaning grass, strewed before church doors on saint's days in north Europe, burned as an incense for spiritual purposes or woven into baskets in North America, its fragrance resulting from the presence of the chemical coumarin, $C_9H_60_2$. . . .

Little enlightenment here about the sanctity of grass nor about how, in so many ways, it relates to other life-forms of Earth and to our own wholeness and health. The factual knowledge that serves so well the workaday world serves also to disenchant and distance us from the wild universe that exists both around and within us.

Aldous Huxley wrote about perceiving the marvels of common things by "cleansing the doors of perception," a poetic phrase borrowed from Blake (whose doors must have been of glass).[1] By the time we reach adulthood the doors/windows of the mind are rendered semi-opaque, encrusted with layers of traditional ideas that obstruct the clear light of reality. Mature perceptions of the world are plastered over with learned conceptions. A similar idea is expressed by those who have praised the lucid "child-like mind," because childhood perceptions of the surrounding world are less obstructed by cultural baggage. The pure joy exhibited by children on seeing and naming things must surely be the basis for the biblical story of the first child, Adam, naming the animals. Heaven lies about us in our infancy.

Suppose that "cleansing the doors of perception" were widely accepted as a worthwhile goal, derived from the proposition that the way we have been taught to perceive reality is flawed. Again, suppose that many sincerely wish to regain a child-like appreciation of simply being. The weight of tradition, as summarized in Huxley's book, *The Perennial Philosophy*, suggests turning inward.[2] Prayer, meditation, and contemplation are recommended to bring peace and comfort to the mind. This may be a necessary step for many, though only a preliminary step because it is personal and individualistic, not impersonal and ecologic. The goal of "transcending one's self-conscious selfhood" is good, but the practice is fundamentally homocentric. It proposes to bridge the rift between such dualities as Man/God and Body/Spirit and, after that, to deal with the question, "What about inter-human relations?" While healing the conscience, and perhaps encouraging altruism, the inner light cannot illuminate the truth of the People/Earth relationship whose disclosure requires a turning outward.

Finding Our Way Back

Immediate perception reveals a sensual material world; all else is gossamer spun by minds inclined to disparage the primary evidence of the senses. Those of us who desire a present paradise, a garden-Earth, rather than escape to spiritual realms or to other planets, must be believing *materialists* in the best sense of that misused word. The Nature from which we came has a substantial body that each of us appreciates as sights, sounds, odours, tastes and feelings—all creatively fused in the wide world and in subsidiary bodies such as our own. Therefore the true materialist is a lover of *this world*, a believer in the supreme value of *this world*, confident that the spiritual is derivative from the aesthetics of the material. Our psychological responses to beauty are immediate, emotional, anchored in daily experience, strongly Earth-centred. As a guide to what is ecologically healthy and unhealthy, we must polish, refine, and trust the sense of beauty for Earthly things, buttressed and supported by understanding and sympathy that disclose our extraordinary "mundane" origins and relationships.

Once our thoughts are straight, once we recognize the importance of Earth relative to ourselves, we will honour the planet as vital source rather than resource. Aware of the cultural obscurants that have blinded us to our ecological connections, we can begin to shuck them off and shine up the windows of perception. More positively, we can seek ways to forge again the ancient connections with surrounding Nature.

How can we assuage the deep-seated yearning for a sense of place? How can we find a mooring in Earth to steady us in times of economic upheaval, global environmental crisis and mass culture? We should at least entertain the unpopular idea of doing less, concentrating on just *being* in the world. A few centuries ago, the ideal person was a do-nothing contemplative like Catherine of Genoa or Meister Eckhart, rather than a hard-working, go-getting entrepreneur. Now we censure indolence, except when flaunted as the badge of affluence. Should we not think of indolence as a virtue in Bertrand Russell's sense; that is, of moving the fewest possible material things from place to place? Work, he said, should be minimized, not considered a duty, and idleness should be universally accessible.[3] Ecological contemplation could be the new mysticism. Music and song and poetry would still be welcome. Taking a leaf from the plant kingdom, the original Greens, the "do-nothing" stance in the sunshine is the worthy *way* (Tao) of the world.

How foolish to laud a work-ethic that keeps people nose-to-the-grindstone, busy making and doing far beyond their needs. Here is a monumental mistake. We ought to do less, simply because much of so-called "productive labour" is destructive: it consumes resources, encourages over-population, creates garbage and weakens the Earth-source. Wealth is much more a problem than poverty. By working less we could free up time for the more worthy task of self-harmonizing with the surrounding reality.

Living Lightly

Ideally, everyone should live on the open prairies or in forests, beside mountains and lakes, where every day fresh breezes blow over landscapes of grass or trees, clouds, rocks and water—daily reminders of the real world mostly foresworn. The bad news is that demographers predict more and more of us will live in cities and, simply put, the city is an unhealthy place for those who want to come home at least once before they die.

In the city it is possible to be a God-fearing person, a humane person, a kind-to-your-neighbours person and at the same time be a bad materialist, selfishly pursuing personal possessions. In the city it is difficult to be a good materialist, a good ecologue, rapt in appreciation of

Earth's eternal creativity, feeling a rapturous part of it. Cities have their undoubted attractions, but the context is too much concrete, too many machines, too much exhaust, too much noise, too much light, too many people, and not enough peace and quiet, not enough sensory variety, not enough birds and trees, water and grass, northern lights and rainbows.

Are you stuck in the city? Don't give up the search for Holy Grass near your doorstep. Nurture the beautiful, plant bulbs for spring flowers, tend a garden, make music, make love, go out in the rain in summer, roll in the snow in winter. Take frequent holidays from the depressing daily news and the talking heads that convey it. Nurture humour. As the wise old mycologist reminds us, from time to time we need to break out of the adult mould and be fungis and fungals. Join the Natural History Society, the Parks and Wilderness Society, and work hard to get very large natural areas preserved against any kinds of "development," secure in the knowledge that in ninety nine cases out of a hundred further alterations to Earth's ecosystems are misguided—as the haphazard happenings in the Bow Valley Corridor west of Calgary and in the Fraser Valley Corridor east of Vancouver clearly show.

At the social/political level, green your political party as you green your city or town. If we let our imaginations roam, perhaps we can recreate the non-human natural, the splendor and the silence, near where we live. Why not more front lawns, boulevards and parks simulating and celebrating flowery native prairies and woodlands? Why not at least brief dimming of street lights on clear winter evenings so children can see the grandeur of the night sky? Why not a few near-silent hours on sunny afternoons, vanquishing, if not obliterating, the noise pollution of internal combustion engines? Is commerce never to give us any rest?

Most people can occasionally escape from the city, a happy fact. We should get away as often as possible, though the standard white-knuckle drive in a private automobile down a four-lane highway needs rethinking.

What, then, does the city-escapee do? For children the answer is easy. Simple exposure to Nature, without didactic teaching, is all they need to tune themselves to Earth's creative aura and bond emotionally to it. Just turn them loose in a garden or any wild spot. For adults the way is harder. Take a break from cultural history, from *People Magazine*, and study your own natural history and that of the wild world. Learn about the ecoregion in which you live: its history, its geology, its climate, soils, flora and fauna. Go in the sun and the wind to contemplate non-humanized vistas, feel their loveliness, their charm and your bone-deep relationships to them.

The Holy Grass, the Sacred Ground, is near your doorstep and out beyond . . . waiting for you.

References

[1] Huxley, Aldous, *The Doors of Perception* and *Heaven and Hell* (New York and Toronto: Harper & Row Publishers, 1954).

[2] Huxley, Aldous, *The Perennial Philosophy* (New York and London: Harper & Row Publishers, 1945).

[3] Russell, Bertrand, *In Praise of Idleness and other Essays* (New York: Allen & Unwin, 1935).

CULTURE AND CREATIVITY

Reprinted with permission from *The Structurist*, No. 33/34, 1993–1994. University of Saskatchewan, Saskatoon, pp 66–70. Edited for this collection.

The universe is creative. In the fifteen billion years since the Big Bang it has evoked cycles of star making, constellations, solar systems, planets, ecosystems, organisms, and humans possessed with a conscious spark of that creativity. Why does the Earth-born human not harmonize more closely with the creativity of universe and Earth? The harmony seems vivid and enchanting in childhood when the mind is ingenuous and the sense of wonder intense. Perhaps we have misunderstood both the authenticity of childhood's imaginative vision and the limitations of the "grown-up" wordview that displaces it.

In adulthood, from time to time, many still experience nostalgic glimmerings of that youthful reality, seeking it again in summer and winter recreational outings, felt as a vague yearning for the world of nature and for the spirit—"anima" and "animus"—with which it is imbued. To recapture the primal enchantment starting from where we are may seem almost impossible. Nevertheless this is a high and worthy goal: to find and return to the springs of creativity, to be once again attuned to Earth.

Spirit of Place

We are a theorizing race, seeking the overarching generalization, the explanation that knits together in some comprehensible way the bits and pieces of our fragmented experiences. One method is scientific; for example, counting flower parts to classify and thus make more understandable the botanical world of plants (Aha, the Saskatoon bush is a member of the rose family!). But there is also the way of artists when, as intuitive "joiners" and "articulators," they bring things together in imaginative and affective ways.

By this argument, though they may object, scientists and artists are religious people, for the fundamental meaning of "re-ligio" is to bind or tie together again. The only sin is separation. Artists and scientists, like the truly religious, creatively seek unifying experiences that make aesthetic and intellectual sense within Earth's mystifying milieu of sky, land, water and organisms.

To make sense of the universe and self is a potent urge. Victor Frankl, philosopher and psychologist, asserted that people are driven not by Freudian sex, or Adlerian power, or Jungian strife between the unconscious and the conscious, but by the ethical search for meaning in their lives.[1] Rooted physically and mentally as we are in the planet, the meaning we seek must also have planetary roots. The hypothesis is reasonable that Earth continues to speak to us not only through the innate intelligence of each body/mind but also, to the sensorily aware, through the outer "spirit of world and place."

People in today's secular societies shy away from the word "spirit" because of its association with outmoded theological doctrines and, in a science-ruled world, because of its association with subjective and unmeasurable aspects of experience. But subjective and unmeasurable aspects of experience do exist importantly, on the valued side of living. Qualitative thoughts and feelings cannot be ignored—we need words for them.

"Spirit" from the Latin *spiritus* meaning breath, vigor, life, conveys authentic values. So does that other Latin word *anima*—the breath, soul, or life principle—from which came "animated" and "animal" as well as the phrase *anima mundi* (feminine soul of the world) that expressed Plato's concept of the world as organically alive.

"Spirit," then, is the energetic and animating principle that we ourselves possess and of which we are aware by simply experiencing our own aliveness, conscious of our immaterial thoughts and feelings. The idea of spirit as energetic also provides a link with science through the discoveries of physics that everything from subatomic quarks to the galactic universe is energy and energy's expression. Matter has been dematerialized; only energy/spirit remains!

The Yogic tradition identifies energy nodes and flow patterns in the human body, drawing their power from Earth and the Universe. Here is an ancient endorsement of today's realization that people derive what energy they possess from sun and Earth via the landscape/waterscape ecosystems within and around which they dwell. Like all else, we of the human species pulse with energy, with animation, with "spirit," and this is one of the great scientific, artistic and religious rediscoveries of the last one hundred years.

The idea that energetic world-spirit streams and gathers in special landscapes and waterscapes was taken for granted in earlier cultures. The phrase "spirit of place" translates the Latin *genius loci*, a reminder that long ago the Romans had their set-aside "natural areas": special groves and streams where dwelt the nymphs, the dryads and the naiads. In such sacred places people felt Earth's mystical, numinous power. Long before the Roman Empire the Chinese practiced the art of Feng-Shui (wind-water), a form of geomancy whose purpose was "to divine the Earth spirit."[2] They aimed to trace the flows of energy circulating endlessly in Earth's veins and vessels, striving to harmonize their built environments with streams and currents of the cosmic breath. Thus they fitted the sites of their towns, temples and tombs to land and water sensitively and aesthetically, taking clues from an intuitive sense of landscape ecology.

In both Western and Eastern traditions, those seeking spiritual enlightenment—the illumination of their lives by "spirit"—have usually searched within, eyes closed, by introspective meditation and contemplation. The *genius loci* concept balances the inward with outward attention, encouraging the search for "spirit" in nature. In the words of Tsung Ping, Chinese landscape painter of the fourth century AD:

> Landscapes have a material existence, and yet reach into a spiritual domain. The wild beauty of their form, the peaks and precipices rising sheer and high, the cloudy forest lying dense and vast, have brought to the wise and virtuous recluses of the past an unending pleasure, a joy which is of the soul and of the soul alone. One approach to the Tao (the Way) is by inward concentration alone. Another, almost the same, is through the beauty of mountains and water.[3]

In suitable natural places, in moods of openness to the Earth, people still experience the *genius loci* with its affective drawing power. This apprehension may come unexpectedly—from simply walking in a majestic forest, seeing mountains in the moonlight, hearing birdsong at dawn, smelling the fragrance of wild syringa or lilacs on a June day. Why is it no longer a power in the public mind, connecting ethical attitudes to Earth? One reason may be that fewer and fewer people have opportunities to experience an ever-decreasing number of natural ecosystems. Further, the major beliefs and technologies of industrial society are so lacking in ecological sanity, so far removed from the realities of non-urban landscapes, that people have limited access to Earth-spirit insights. These, the rare inspiration of adulthood, are fortunately more accessible in childhood.

The Enraptured Child

The romantic poets drew heavily on their nature experiences as children. The childhood vision was recognized as direct and clear, apprehending a primal reality later obscured by age. Wordsworth wrote of his youth:

> For nature then . . .
> To me was all in all. I cannot paint
> What then I was. The sounding cataract
> Haunted me like a passion; the tall rock,
> The mountain, and the deep and gloomy wood,
> Their colors and their forms, were then to me
> An appetite; a feeling and a love,
> That had no need of a remoter charm,
> By thought supplied, nor any interest
> Unborrowed from the eye.[4]

Theodore Roszak added an ecological and evolutionary dimension to the same theme. Because we are creatures of Earth, made from its dust, air and water, evolved in its forests and savannas, we are naturally adapted to its rhythmic processes. Because our species is one of the twig tips of a life-tree extending back to the beginnings of time, we retain in our body/minds the ancient instinctive wisdom of all ancestral forms. The repository of ecological/evolutionary knowledge is the unconscious—the id of Freud, the collective unconscious of Jung—and through it Gaia speaks to us still, especially in childhood. Roszak proposed that the developing child's psyche recapitulates the psychic history of the race. Hence, given exposure to nature, the child's pre-egotistical vision is of self and the world as one. Children possess a rare noetic talent, he wrote, and their innocence endows them with a purity of perception. They greet the natural world with an instinctive animist response; to them it is alive and personal. In the lucidity of their experience something of the old sacramental vision of nature is reborn.[5]

This being so, childhood psyches are stunted if their contacts with the natural world are prevented, as in the inner city, and this retardation carries over into adult life. Edith Cobb provided evidence that important aspects of childhood experience remain in memory as a psychophysical force, an elan, that fosters the potency to perceive creatively and inventively in later life. Her studies of the child in nature, culture and society revealed, she wrote, that there is a special period, the little-understood, pre-pubertal, halcyon, middle age of childhood—approximately from age five or six to eleven or twelve, between the strivings of animal infancy and the storms of adolescence—when the natural world is experienced in a highly evocative way, producing in the child a sense of some profound continuity with natural processes, and presenting overt evidence of a biological basis for intuition.[6]

Thus the genius of human creativity derives at least in part from childhood experiences of the *genius loci*; that is, from a living relationship between person and place. A child's self-exploration does not produce a knowledge of the world, she said, but rather immediate knowledge of the natural world produces the child's real self. Notice that the injunction to "know thyself" has ecological overtones. The unmediated visions of childhood—engaging simple things like trees, gardens and ponds—may inspire the imaginative spirit of the adult.

Unless children are given opportunities to "bond with nature" and fulfill their ecological needs, they are unlikely in later life to feel drawn to their source. Paul Shepard traced modern environmental destruction to the failure of society to provide the kind of natural environment and experience necessary for a young primate's development.[7] The first cultural misstep in human history, he said, was the discovery of agriculture, a turning away from companionship with unsimplified, undomesticated nature. In a similar vein, Ted Williams asked the question, "Why does Johnny shoot stop signs?" Youngsters imprinted on machines and other inorganic artifacts cannot be expected to empathize with the wider world. According to him the same deficiency is expressed in whole nations: "the Scandanavians band birds; the Italians eat them." Japan, the very model of a modern society both commercially and educationally, is one of the most ecologically illiterate.[8] As yet there is no country prescribing "nature imprinting" for all of its children, devoting as much curriculum and exposure to natural history as to cultural history.

Ecological Insanity, Social Madness

Two theories of human nature confront one another. The first diagnoses humanity as basically a wicked lot, conceived in sin and thereafter badly in need of redemption and salvation. Certain religions propagate this view, and it has been championed by secular thinkers, such as Sigmund Freud. He theorized that people are largely run by their unconscious minds: seething cauldrons of sexual energy forever threatening to blow the cover of respectability, the violent Mr. Hyde behind the proper Dr. Jekyll. The instinctual demon within is controlled by society's rules and regulations, its laws and expectations. In the best circumstances the libidinous energy is sublimated, channeled into constructive paths of good citizenship and good works. In the worst cases, repression of the passions produces anxiety, melancholia, anti-social behavior and—lightening the load according to Freud—artistic achievements![9] Aside from artists, whose madness up to a point is socially acceptable, those unable to conform to the "reasonable" dictates of culture are the losers, the neurotic, the lawbreakers.

The opposing theory posits people as fundamentally good. The newborn babe, the child, is natural, outgoing, imbued with a sense of wonder, infatuated with animals and plants, no sex hang-ups, no prejudices, completely at home in the world. This is the basic tenet of most post-Freudian schools of psychology and psychiatry, asserting the essential health and innocence of human nature. It is also the message of numerous New Age therapy groups whose advertisements for authentic personal development and self-actualization ornament the billboards of all large towns and cities.

In the eighteenth century, Jean Jaques Rousseau set forth the second, the optimistic viewpoint, in his influential book *Emile*, the source of many of the principles of modern education.[10] At birth, said Rousseau, the child is natural and unspoiled. Education should do nothing but give the child an opportunity to develop its natural gifts unhampered, while shielding it from the corrupting influences of society. According to this idea the instinctive minds of young humans are tuned to the natural world that is experienced first as the loving mother and then as a landscape of wonder. But society, through the instrument of socializing parents and teachers, turns the child away from its initial animistic empathy with Earth. Encouragement of the "I–it" relationship severs psychic rootedness, setting the stage for conflict between the conscious and unconscious aspects of mind, generating anxiety and neurosis. In short, Western culture—more and more city-based, further and further removed from any grounding in Earth-wisdom—systematically drives its citizens insane.

A society that renders its citizens mad must itself be mad, a thesis advanced half a century ago by Eric Fromm who foresaw a robotic world of machines that acted like people, and deranged people that acted like machines.[11] Today the primary evidence of madness in Western culture is the war waged against Earth, the very environment that sustains it. What but madness condones amassing human wealth by destroying nature's wealth, extinguishing whole land and ocean ecosystems along with their plants and animals, degrading air, water and soil, reversing evolutionary trends by simplifying nature's complexity, risking entire webs of biodiversity by elevating the monoculture of one species to the highest good, setting no limits to population numbers in a fixed Earth space?

The Second World War never ended in 1945; it simply continued as a battle against nature using the same weapons: deadly chemicals sprayed from airplanes on fields and forests, attacks on the land with caterpillar-tread tanks and other aggressive machinery, the diversion at last count of about 40 per cent of the net photosynthetic productivity of the world's terrestrial ecosystems to human uses with consequent starvation of millions of companion organisms. The results are collapsing fisheries and forests, spreading desertification, industrial pollution, acid rain, ozone holes. The casualties of a "take no prisoners" policy are Earth's supportive matrix of air-soil-water, dependent plants and animals, and, lately, people as ecological refugees.

Mind-Forged Manicles

Suppose that each person is born with a capacity to live lightly and symbiotically on Earth: a natural Earthling. But soon the surrounding demented culture, mostly invented by the male sex over the last few hundred years, shapes them into anxious consuming and polluting monsters. How is this done? The poet William Blake suggested "mind-forged manicles"—in other words, the commonly held, taken-for-granted ideas embedded in language. Developed into thought-systems, the manicles are ideologies, faiths, ruling theories or paradigms. Unconsciously and unquestioning, the child soaks up these cultural belief-systems by osmosis from parents, friends, teachers, books, TV and other media.

Societies function on the basis of the cultural myths they espouse. Belief systems channel and constrain the thoughts and actions of all people. Paradigms lead, facts follow. Applying these ideas to the scientific community, Kuhn suggested that most research workers in any discipline do not question the popular theories within which they labour. They accept on faith the current framework of beliefs as to how nature and society function, content to tackle the puzzles that the framework suggests are important. Eventually, said Kuhn, this "normal science" runs into stubborn odd-ball problems or anomalies that cannot be solved within the taken-for-granted paradigm. The field or discipline is "stuck," awaiting an improved theory—from an Einstein or a Darwin—followed by a system shake-up and a "reconstruction of group commitments."[12]

Today the urban-industrial belief-system is sticking, or perhaps in the sense of falling apart, getting unstuck, and the time is right for a reconstruction of social commitments. The ruling theory is unbalanced and irrational. If society continues to live by it—lip service to "sustainability", "good stewardship" and "respect for Earth" notwithstanding—the current destruction of Earth's biodiversity and cultural diversity will accelerate to a disastrous end. For the God-word of the current ruling paradigm is growth—economic growth, population growth, agricultural production growth, city growth, consumer and consumption growth. Based solely on human wants, it shows not the slightest understanding of the close psychological and physical relationships between the human race and Earth's ancient ecosystems. It recognizes no limits to the expropriation of Earth's wealth. Such a ruling paradigm, forged by our ecologically naive ancestors, is clearly false and foolish.

Ecological ignorance today is inexcusable. The truth is we are Earthlings in body, mind and spirit. The very fact of existence means that in the past our ancestors successfully adapted ourselves to the realities of Earth life. Today the need is urgent for a paradigm that again harmonizes us with the cosmos, accepting as right and good an affectionate, child-like outlook inoffensive to Earth and all its parts, a worldview that recognizes not only the spark of creativity in people but also the fountainhead of that creativity in Nature, manifest in the spirit of place.

References

[1] Frankl, Viktor E., *The Unheard Cry for Meaning* (Toronto and New York: Washington Square Press Publication of Pocket Books, Simon & Schuster, Inc., 1978).

[2] Anderson, Richard Feather, "Geomancy: Divining Earth Energies and Rhythms," *The Trumpeter* Vol. 2 No.4 (Victoria: LightStar Press, 1985) 13-17.

[3] Tsung Ping, fourth century AD Chinese landscape painter, quoted by Tom Bender, *Environmental Design Primer* (New York: Schocken Books, 1976) 145.

[4] Wordsworth, William, "Lines Composed a Few Miles Above Tintern Abbey" *The Norton Anthology of Poetry* ed. Arthur M. Eastman (New York: W.W. Norton & Company, Inc., 1970) 242-243.

[5] Roszak, Theodore, *The Voice of the Earth* (New York and Toronto: Simon & Schuster, 1992).

[6] Cobb, Edith, "The Ecology of Imagination in Childhood," *The Subversive Science, Essays Toward and Ecology of Man*, eds. Paul Shepard and Daniel McKinley (New York: Houghton Mifflin Company, 1969) 122-132.

[7] Shepard, Paul, *Nature and Madness* (San Francisco: Sierra Club Books, 1982).

[8] Williams, Ted, "Why Johnny Shoots Stop Signs," *Audubon* Vol. 90 No. 5 (September 1988) 112-121.

[9] Matson, Katinka, "Sigmund Freud," *The Psychology Today: Omnibook of Personal Development* (New York: William Morrow and Company, Inc., 1977) 205-210.

[10] Rousseau, Jean Jaques, *Emile* (New York and Toronto: Little Brown, 1993).

[11] Fromm, Eric, *The Sane Society* (New York: Holt, Rinehart & Winston, 1955).

[12] Kuhn, Thomas S., *The Structure of Scientific Revolutions*, Second Edition (Chicago: The University of Chicago Press, 1970).

BOREAL WILDERNESS

Bigbeautifultractsofnaturalforestland,suchastheborealDore-Smoothstone Lakes area in northwest Saskatchewan, are essential components of this miraculous planet. Such wildernesses and semi-wildernesses possess their own inherent values quite apart from those of humankind—the leading alien species and invasive forest pest. Traditionally we have viewed forest ecosystems through the distorting glasses of economists that reduce them to one familiar form: "natural resource" wealth to be creamed off by industry, with a split for government.

So it is that Canada's boreal forest has earned the title, "Lumberyard of the World." Apart from their money-making lumber and pulp, all else that northern forests harmoniously embody and nurture in the way of diversity, watershed protection, wildlife habitat, and natural landscape beauty has been deemed of little value.

This myopic perspective needs to be changed, and change is on the way. As with all shifts in public perception, the appropriate outlook is heralded by better vision, by the encouraging words of a few radical politicians (who may be right for the wrong reasons), and especially by citizen-group action on the ground level.

The new vision perceives humankind as a cooperating partner with Earth rather than its ruler. As junior partner, people need to understand the Gaian enterprise and our ecological duties to ensure its continuing prosperity. Forest ecosystems are home not only to humans but also to countless other life-forms, all evolving together in various mutualistic and competitive ways under Earth's guidance. Human societies foster their own sustainability by fostering ecosystem integrity, avoiding the simplifying practices that reduce forestland diversity. Opportunities still exist to reject the industrial model that has turned most of the planet's natural polycultures—deciduous forests and grasslands—into agricultural, single-species plantations.

In 1995 the Canadian Council of Forestry Ministers approved as national policy a number of criteria judged necessary for sustainable forest management. The first four—the basis and support for the others—say nothing about turning trees into two-by-fours and Kleenex but, surprisingly and commendably, much about protecting the forest landscape by assuring:

1. Conservation of biological diversity.
2. Maintenance and enhancement of forest ecosystem condition and productivity.
3. Conservation of soil and water resources.
4. Maintenance of forest ecosystem contributions to global ecological cycles.

In practice, all four criteria require sensitive ecological land-use planning that is far-sighted and spatially inclusive. The national policy implies that tracts of protection-forests, wildlife corridors, and areas of historical-spiritual significance will be zoned out of intensive commercial use. Timber-producing zones, not necessarily the largest proportion of the landscape, will be planned and managed in ways that over the long term perpetuate the natural composition and structure of the mosaic boreal forest. Large natural and semi-natural tracts will preserve normal ecosystem constituents and processes, moderating the regional effects of resource extraction.

Despite these fine words, the majority of those in government and industry—geared to milking nature's cash cows—still need to be educated and persuaded. The dynamic for change lies with groups of citizens, organized in non-government organizations (NGOs), motivated by the ecological world view that places *Homo sapiens* as a cooperative and sustaining part of Earth, not its CEO.

One obstacle to sustainable forest management is the delegation of provincial authority over forests to company licensees. The effect, in addition to conveying rights to timber, is to privatize the responsibility for environmental protection—the responsibility for protecting biodiversity, water, soils, plant-and-animal habitat, and aesthetically pleasing landscapes—without clear authority, incentives, or resources to cover the costs. Small wonder that forest licensees systematically under-invest in environmental protection.

Governments must either legislate rules compelling forest-product companies to use a part of their profits to protect the non-timber values of forestland, or else do the job themselves. When government and industry fall short of practicing the kind of management that will assure the survival of natural forest ecosystems, then the activities of NGOs such as the Canadian Parks and Wilderness Association become crucial. Organized in various watershed-protection, wilderness-protection, and biodiversity-protection groups, lay-people as researchers and guardians should be welcomed as allies by government. They can assist in promoting ecologically sound forestry practices while detecting and defending against those that are harmful.

Protection of the Dore and Smoothstone Lakes Wilderness Area is a trail-blazing example of citizen participation in caring for an attractive and delicate part of the boreal forest. Everywhere in North America the example should be followed. Old-growth forests deserve first attention for they are irreplaceable. The proposition that, once logged, we know how to bring them back is a falsehood.

TRADITIONAL ECOLOGICAL KNOWLEDGE

Of what relevance to current industrial civilization are indigenous cultures, those islands of Aboriginal society that still persist here and there in the turbulent sea of world trade and commerce? Does their traditional ecological knowledge (TEK) offer useful values and lessons for those of us seeking harmonious and sustainable ways of being on Earth?

My thesis is that their relevance is limited because of a fundamental cultural gap. The worldview to which we have been born and raised, ecologically deficient though it is, cannot simply be exchanged for any one of the many others developed by foraging, Nature-revering societies. We must find our own way with a new faith allied to, or at least not contradicted by, the best salvageable parts of the modern and post-modern culture that is the Western inheritance. This is not to discourage Aboriginal societies in their specific belief systems, especially in those aspects that embody ecological wisdom, for rejection of the arrogant Western missionary mind-set that has harmed them as a people in the past is long overdue.

Anthropologists and sociologists are interested in the flexible ways diverse cultures adapt to the various physical and biological landscapes of Earth. One modern approach is the study of historical ecology, tracing patterns of change in cultures as they evolve. The presumption is that "cultures"—here meaning the foundational beliefs, faiths and values of societies—are subject to historic selection, responding and adapting both to external forces and to internal dynamics or, if failing to do so, disappearing. This is the story of past civilizations.

External forces that shape Aboriginal cultures vary according to geographic place. Such forces are expressions of the natural processes of the ecosystems in which the tribes live: the conditions of climate, soils, water, plants and animals, the periodic incidence of large-scale environmental change (especially climatic change), and human interactions as cooperation and/or competition with neighbouring societies.

Internal forces, the focus of interest in this article, are the historically developed, slow-changing beliefs that determine the practical engagement of people in the world. The foundations of every tribe's sustainable ways of living, and its long-term survival, lie in its belief-system, its cultural faith-narrative.

The question is whether, in a world rapidly being overrun by Western Civilization (aka global capitalism, global corporatism, global industrialism), the study of any but our own "practical engagement in the world" has relevance to the current scene. Does the historical ecology of various indigenous cultures around the world illuminate our future path? Can we transpose aspects of their accommodation to the Earth into our quite different culture?

Affirmative answers point out that much can be learned from those indigenous cultures that exemplify sustainability over centuries. They can teach the fundamentals of living with one another and with Earth in ways that are relation-based rather than consumption-based, responsibility-based rather than rights-based, communal rather than individualistic. We examine these Aboriginal cultures and marvel at their ways-of-living that often seem so wholesome compared to our own. Here apparently is indigenous peoples' knowledge that can be borrowed and used.

We of the "Western Civilization" comprehend most readily the visible aspects of other cultures, particularly the activities of daily living whereby tribal people interact with each other and with the lands that enclose them. Anthropological studies detail these observable features: the small population in a close-knit community, the binding rituals, methods of settling disputes, the foraging habits, use of medicinal plants, organic agriculture, small but sophisticated technology, reliance on solar power, taboos on over-hunting, and so on. Such relatively easy-to-understand habits demonstrate, some say, "the ecological way to go."

Indigenous cultures do offer clues and transferable facts, especially for such land-based activities as management of wildlife and the identification and use of herbal foods and medicines. These are useful in our urbanized culture but marginal, because few of us live in a land-based way. No longer are we alive in our senses, as they are, to Nature as the source of food, clothing, shelter, and spirituality. Placeless in the sense that all towns and cities are much the same, we search indigenous knowledge with the expectation that it, like our own "know-how," is factual and abstract ("scientific" in other words), readily transferable by the printed page to any culture in any part of Earth. Such is not the case.

Indigenous knowledge is an outgrowth of close historical and ecological relationships to particular geographic places; it represents the lived experience of foragers, gatherers, hunters, and simple agriculturists. It is a whole set of concepts, beliefs, and stories—a cultural construct—dealing with processes and activities that relate people to the particular parts of Earth with which they and their ancestors have been in close touch—that is, to the sensed natural world and its presumed invisible counterpart.

According to McGregor, "Capturing a single aspect of traditional knowledge is difficult. Traditional knowledge is holistic and cannot be separated out from the people. It cannot be compartmentalized like Western scientific knowledge. This means that, at its most fundamental level, one cannot ever really 'acquire' or 'learn' TEK without having undergone the experiences originally involved in doing so." You must, "Accept our knowledge as an entirely different way of generating knowledge, not simply useful content to be inserted in Western scientific frameworks," wrote Simpson.

The essential soul of tribal people—what holds them together and makes their life-style sustainable—is invisible, relatively inaccessible and strange to us. Even if a committed student spends long years with a tribe or close-knit ethnic group, empathically exploring and then explaining for us the people's cosmology and fundamental beliefs about themselves in the world they occupy, the relevance of the latter to improving or redirecting our industrial society seems minimal. What we learn is curious and alien, as shown by Williamson in his years of learning the language and ways of the Inuit. Indigenous knowledge derives from place-specific oral traditions, from languages rooted in the land ("land-guages"). Indigenous knowledge does not conform to the linear logic of the printed word. Therefore it does not "fit" what we know of biology, ecology, psychology, evolution, geology, and astrology, nor does it jibe with our understanding of sociology, economics, and politics.

Every effort should be made to preserve cultural diversity globally, and not just because much of academic interest can be learned from remnant indigenous cultures. One sound truth they teach is that way-of-living is intrinsically bound up with fundamental beliefs, with faith often described as "spiritual" (though cultural in the deepest sense) that gives each tribe member a sense of belonging, as well as confirmation that the tribal way of being-in-the-world is good. This sense of universal relatedness we moderns lack and, in fact, the very mention of "spiritual" with its overtones of religiosity closes the minds of many. But "spiritual" need not mean transcendental in an other-worldly sense. Fritjof Capra has rescued the word from theological mists and New Age fogginess, defining it as a sense of belonging, a sense of connectedness to the cosmos. "Ecological awareness," he notes, "is spiritual in its deepest essence."

Many people accept and support much of what tribal peoples do in the name of their cultures, admiring and perhaps imitating through the voluntary simplicity movement some parts of their sustainable life styles. Our difficulty is that we cannot understand nor accept the deeply rooted religious or philosophical faiths that support them in living the way they do. Without that radical, binding glue, the communes and other utopian tribal experiments we attempt, even when "love-based," soon fall apart. The utopian experiments have been tried again and again in the Western world and, after a generation or two at most, all have failed.

Knowing that fundamental beliefs and way-of-living are unitary in culture, we who are swathed in Western culture cannot hope to borrow even from the most enlightened indigenous cultures either their fundamental beliefs or their way-of-living, let alone both together. To switch Western culture from its present track to a saving ecopolitical route means finding a new and compelling belief-system to redirect our way-of-living. It must be a vital outgrowth from the twenty-five-hundred-year-old Western culture into which we have been born and of which we are the heirs.

It seems to me that the most promising belief-system is Ecocentrism, defined as a value-shift from *Homo sapiens* to planet Earth, to the Ecosphere. An ecological rationale backs the value-shift. All organisms are composed of Earth's elements; all are evolved from Earth, sustained by Earth. Thus Earth, not organism, is the metaphor for Life. Earth not humanity is the creativity-centre, the life-centre. Human societies and their languages, their sense of aesthetics, their physical and mental health, are outgrowths and creations of Earth's ecosystems. Earth is the whole of which people are subservient parts. Such a fundamental ecological philosophy gives human awareness and sensitivity an enfolding, material focus.

Whatever the new ecological way, it must be advanced through some form of community and communal living. Traditionally the word "community" has meant a gathering of people, a one-species society, with secondary attention to context and to other creatures. Ecocentrism suggests a broader and more inclusive interpretation. The Ecosphere is central and it constitutes the largest "world community." Its components are geographic communities *as places on Earth*: the geoecosystems, recognized at various size scales from the regional to the local, to which humans must sooner or later accommodate themselves. Such a definition makes "community" a fully functional living entity, rather than an exclusive aggregation of human beings. In this sense a minimal Earth community can exist without human occupancy, or as one person or family at home with and caring for a piece of Earth and its companion organisms.

Ecocentrism is not an argument that all organisms have equivalent value. It is not an anti-human argument, nor a rejection of those whose focus is a commendable approach to social justice and gender equality. It does not deny that myriad important homocentric problems exist. But it stands aside from these smaller, short-term issues in order to consider ecological reality. Reflecting on the ecological status of all things organic and inorganic, it comprehends the ecosphere as a being that transcends in importance any one single species, even the self-named wise one, *Homo sapiens*.

Ecocentrism is a new way of thinking. It proposes an ethic whose reference point is *supra-human*, above and outside humanity. It places the health of the ecosphere before human welfare. It points the way to solving questions that, within humanistic or biocentric frameworks, are virtually unsolvable: the economic growth problem, the population problem, the abortion-choice problem, the technology problem. It gives new and constructive direction to philosophers, economists, scientists, and engineers.

We cannot adopt, nor make our own, any of the Aboriginal "beliefs & ways-of-living" systems. Perhaps by accepting co-existence models and maintaining communication with other cultures we may gradually approach and come closer together in common beliefs in the future (Gayton, 2002). Meanwhile, as each sustainable indigenous culture in its own way demonstrates, we too can learn to be ecologically aware, looking outward rather than inward, seeing ourselves as dependent Earthlings and not the centre, recognizing our partners—twenty-five million other kinds of creatures—joined with us in a yearly whirl 'round the sun.

We can climb down from the self-erected pedestals that have kept our heads in the clouds and our feet off the ground for the past two millennia and show a little humility. We can get "spiritual" in Capra's sense of seeing ourselves as parts of a greater whole. With new outlook/insight we can view clear-eyed this magnificent, miraculous Earth and recognize ourselves as primarily Earthlings.

Such a new stance would do wonders for our way-of-living. In time it could become the world's TEK: a saving indigenous knowledge.

References

Capra, Fritjof, *The Web of Life* (New York: Doubleday, 1996).

Gayton, Don, *Indigenous People's Knowledge and Western Science: a personal view. Linking Indigenous Peoples' Knowledge and Western science in Natural Resource Management.* Conference Proceedings p. 7-9. Southern Interior Extension and Research Society Series 4, 2002. 478 St. Paul Street, Kamloops, BC V2C 2J6

McGregor, Deborah, "Traditional ecological knowledge and the two-row wampum" *Biodiversity* Vol. 3 No. 3 (2002) 8-9.

Simpson, Leanne, "Commit to and respect Indigenous Knowledge—before it's too late" *Biodiversity* Vol. 3 No. 3 (2002) 32.

Williamson, Robert G, "The Arctic Habitat and the Integrated Self" *Landscapes of the Heart: Narratives of Nature and Self*, eds. M. Aleksiuk and T. Nelson, (Edmonton: NeWest Press, 2002) 167-204.

WATER AND FORESTS

The oldest epic story, five or six thousand years old, has been found in various versions written on clay tablets in the ruins of the ancient city of Nineveh in present-day Iraq. It tells of King Gilgamesh, ruler of the city of Uruk in southern Mesopotamia. There on the floodplains of the Tigris and Euphrates Rivers early civilizations prospered. To build up his city and its walls, Gilgamesh and his trusty companion, the wild man Enkidu, set out to log the cedar forest in the mountains that border the eastern and northern sides of the agricultural lands known as the "Fertile Crescent." Assisted by the male Sun-god they captured Humbaba the guardian of the forest, chopped off his head, and then proceeded to clearcut the cedar forest. Once logging began, according to the story, "for two miles you could hear the sad song of the cedars."

One version of the Gilgamesh story calls the primeval cedar forest

"The Land of the Living," presumably because "evergreenness" has long been a symbol of immortality. Although they belong to a different genus, our North American cedar's common name, arbor vitae, carries the same symbolic meaning, "Tree of Life." As with the cedars of the "Living-Forest" logged by Gilgamesh, the "Tree-of-Life" name has not prevented the downfall of both the North American eastern cedar and the Western red cedar, neither in BC nor in AD.

Another version of the epic, with a feminine twist, named the mountain cedar forest the *Throne of Ishtar*, and the *Home of the Gods*. Ishtar was the powerful Babylonian Venus, the chief deity in pre-patriarchal society. She had various titles in the ancient world, including (by those who were plugging for a male god) "The Whore of Babylon," but also and more properly "The Queen of Heaven," the "Forgiver of Sins," and "Star of the Sea." The latter name in Latin is "Stella Maris," and from "Maris" (the related French term is "La Mer") came "Mary"—the Virgin clothed in the sea colours of blue and white. Putting it all together, this oldest story of mountain logging pins destruction of the "Land of the Living" on city men backed by a male sky-god, with a poke in the eye for the goddess.

A comment by Dr. Clark Binkley, one-time Dean of Forestry at the University of British Columbia, is relevant to forestry practices not only in mountainous BC but also world-wide:

> With the benefit of history, we now know the rest of the Gilgamesh story: once the forests were gone, civilizations in the region failed and the people suffered from erratic water flows, the loss of forest-dwelling creatures, silted-up harbours, and a shortage of wood for heating, cooking, and construction.[1]

"Silted up harbours" is an understatement because, in the five or six millennia since the saga was written, erosion in the deforested upper reaches of the Tigris and Euphrates Rivers has infilled the head of the Persian Gulf 250 kilometres, with the result that ancient port cities are now one hundred and fifty miles from the sea.

Repeating the Gilgamesh Gesture

The same story of deforestation, erosion, and loss of good water, has been repeated again and again: in the Middle East, in the Mediterranean countries, in the mountainous parts of Europe, and in the Cordilleras of

North and South America. Hence Chateaubriand's lament that forests precede civilization, and deserts follow. "It is a sorry fact of history," said the historian Robert Harrison, "that humans have never ceased re-enacting the gesture of Gilgamesh."

If we could land on the moon, get a distant view of Earth through a telescope, and see the last nine thousand years of our land-use history played like a video—a ninety-minute program at one hundred years a minute—we might understand present difficulties better. In the beginning we would see the spotty clearing of forests along the world's great rivers, in the Middle East and China, as agriculture made its slow start; then the appearance of small cities based on surplus agricultural food, and the steady extension of agricultural land, first along the seasonally flooded or irrigated valley bottoms and then, with domestication of rye, barley, and the desert-grass, wheat, to the un-irrigated, rain-watered plains of Asia and Europe, originally wooded but deforested for cropland. Finally, in the last minute, we would see the ploughing-up of grasslands world-wide for the first time in history, facilitated by John Deere's invention of the steel plough.

Humanity is demonstrating in classic style the principle of positive feedback. More food makes more people, driving the search for more arable land, more food, and more people, then growing cities demanding more land clearing, more intensive agriculture and aquaculture, more forest products, more logging, on and on. In the last few minutes of the video we would see a sudden increase in the spread of agricultural lands and the mushrooming of cities, marked by accelerating deforestation globally. The last thirty seconds would show a veritable onslaught on forests everywhere on Earth, along with much noise and smoke as bigger and bigger machines fed by fossil-fuel hydrocarbons tear into the woods, clearing the land in great swaths, displacing wild animals and chasing most of the wood-workers out.

Water and Civilizations

Historians have identified forty or more past civilizations that arose, flourished for a period, then proved unsustainable and passed away. Evidence suggests that many exhausted themselves by ruining their environments. That's old stuff, some may say. We know better now, and do things differently. After all, we have the science and technology to put men on the moon, so keeping forests and water and soils in a healthy state is a cinch. But is it?

John Ralston Saul suggested that the way people value clean water is a good index of their cultural realism and progress. Any civilization that allows the quality of its water, and its water distribution system, to deteriorate is on its way out. Over the last two thousand years cities have had clean running water three times, when civilizations were at their peaks:

1. The Romans two millennia ago.
2. The Arabs one millennium ago.
3. The West, "until recently."

Saul's comment: "By the standards of the Roman and Islamic empires we are well advanced in our own decline since we have fouled most of our surface sources and are doing the same to our water tables. . . . The sight of millions of Westerners drinking bottled water is a reminder of our disconnection from reality."[2] Saul might have added, the sight of millions of Westerners drinking chlorinated water is not only a reminder of our disconnection from reality but also a reminder of the faith of humans in quick-fix technologies, even those that use known cancer-causing substances to produce "potable" water from what health officials like to call "raw, untreated" water—the high-tech name for the good-tasting liquid we used to drink from lakes, streams, and wells.

Forestlands are the source of the fresh water we enjoy. In temperate regions, with few exceptions, dry grasslands and deserts characterize areas where yearly evaporation is greater than precipitation. Where yearly precipitation is greater than evaporation, forests prevail. In forestlands the excess of precipitation over evaporation runs off to feed wells, springs, creeks and rivers. Thus forests on the eastern slopes of the Rocky Mountains produce rivers that water the dry prairie grasslands. The sources of large rivers such as the Saskatchewan and the Missouri are in forested terrain. Deforesting watersheds contributes to hydrologic disasters, to spring floods and autumn trickles, to soil erosion and the silting in of reservoirs.

Out of the thirsty USA southwest in the early 1960s came a water-diversion scheme called NAWAPA—the North American Water and Power Alliance—proposing to divert water from the Yukon, Peace, and Columbia Basin Rivers into the Rocky Mountain Trench. From this vast reservoir, canals would run south to droughty California, and east to keep the levels of the Great Lakes up, both diversions in the interest of industrial growth. An enthusiastic editorial in the prestigious journal,

Science, in 1965, ended with these words: "It is hoped that Canadians will join us in this great project." We will surely hear again of NAWAPA, or of schemes similar to it, as North American populations and water usage continue to expand. And it may be that the North American Free Trade Agreement will eventually force water diversions and water export from Canada.

Water-transfer schemes should be strenuously resisted. They are technologic fixes, attempts to achieve sustainability of one part of the continent at the expense of other parts, to enrich already wealthy ecosystems by further impoverishing the poorer. Such plans are at odds with the aim of harmonizing humans and their lifestyles, with the resources of the ecological regions where they live—which seems to me the only reasonable long-term definition of *ecological sustainability*. Finding and testing models of sustainable living is humanity's most important task today in all Earth's ecological regions that have been infected by industrial ideas of progress.

Genes and Memes

Knowing something of the long history of non-sustainable cultures that preceded ours today, can *we* find the right track toward living sustainably on Earth? Science pessimists say no. "Biology is destiny," so the argument goes, and human nature regardless of gender is fixed by our genes, which explains why history tells much the same story over and over again, a story of blundering along, repeating earlier mistakes. Perhaps it is human nature to go for the quick take, the fast profit, without much thought of future consequences. According to maverick economist Georgescu-Roegen, humanity's stay on Earth is fated to be a brief but a merry one, like the moth whirling round the flame that soon immolates it.

The more optimistic view, to which most subscribe, is that human nature is alterable. Human actions are the outcome not only of genetic heritage, the *genes*, but also of cultural heritage, the *memes*. In one definition, "memes" are the ideas, concepts, faiths that every child soaks up from parents, teachers, siblings, school chums, the media. Mostly these concepts (the big ones, in modern jargon, called "paradigms") lie concealed under the surface of our thoughts, writings, conversations. They constitute our taken-for-granted culture: the whole network of accepted conventions in discourse that allow us to communicate with one another. They embody the fundamental values that underlie our systems of education, of law, of economics, of politics. Only rarely are they exposed and questioned. We simply accept the cultural norms we are born to, and judge other cultures in comparison to them as "strange" or "uncivilized," "primitive" or "pagan."

"Flexible human nature" expresses the optimistic view of humanity, implying that the damaging things we do to the world and to each other can be corrected if we attend to changing our memes. You are stuck with your genes, but you can change your memes.

The truth about human nature, and what determines it, lies in the interaction of both genes and memes. They are inseparable. Certainly the genetic inheritance from our parents is important: "One day I looked in the mirror and saw that I had become my mother!" Nevertheless the evidence for control of our biological instincts by ideas and faiths is striking. Consider the absurd ideas that the human mind can entertain and act on, some doubtless believed fervently by your nearest neighbours. Think for a moment of the astonishing actions of people who have surrendered their minds to cults, leading in some instances to such anti-biological behaviour as life-long celibacy, self-castration, mass suicide. Remember those who moved to Jerusalem to experience apocalypse first-hand, yearning to be raptured to heaven as the year 2000 dawned? Then consider that every culture is a cult, our own included, riddled with ideas and behavioral patterns that seem ridiculous to other cultures unconscious of their own nonsense.

Fortunately, human behaviour is not set in cement but can be changed. Therefore the conviction that wise guidance in childhood, then education in the values of philosophy and religion, can make us better people by eradicating *bad memes* and replacing them with *good memes*. But what then is a good meme? The answer used to be easy: good memes are the ideas, faiths, philosophies, religions, that solace and profit the human race, especially the ones that help people to live materially better lives. Here is an example of a once-good meme that has turned into a bad one. Such ideas, helpful in the past when we were few in numbers and limited in wants, today spell environmental disaster. Six-and-a-half billion of us, going for ten billion in 2050, all trying to "live materially better and better," are causing acid rain, ozone holes, deforestation, and polluted water, to mention a few of the more obvious of the world's environmental ills.

The philosopher Santayana said that those who know no history are fated to repeat its mistakes. But knowing history will not prevent repeating the same mistakes over and over again if the knowers are motivated by the same human-centred value-memes that caused the mistakes in the first place, whether in ancient Mesopotamia or Periclean Greece, in Mayan Belize, the Easter Islands, or in modern North America.

Ecological Memes

It seems to me that the fundamental flaw of unsustainable civilizations, the common denominator of their unfortunate histories, has been a lack of ecological knowledge and understanding. People have not realized their absolute dependence on what surrounds them: the air, the water, the soil, the plants and animals of Earth. Partly this ignorance has stemmed from philosophies and religions focused on escape from the shortcomings of this Earth (not unlike today's childish NASA-heads). Partly it is a reflection of the human animal's mobility. When conditions become intolerable in one place we have moved somewhere else, yesterday escaping overcrowded Europe and Asia by means deemed legal, today by illegal rusty old ships (in classic cases with the help of a Korean crew), inevitably causing conflicts where we are landed. Also, because the world is flooded with dirt-cheap oil and gas, many of us can stay in place as the environment deteriorates around us, sustaining cities, towns, and villages by trucking in from around the world such necessities as food, liquids, clothing and building materials. This "globalization" is a great danger because it turns attention away from one inevitable need: the caring maintenance by in-dwelling humans of regional and local ecosystems with their water, soils, animals and plants. Without "care of place" all Earth will deteriorate. When times requiring regional and local self-sufficiency return, the means will be much diminished or gone.

With limited ecological understanding (and still today very little of it taught in our educational system), humanity has focused only on itself, believing that people can successfully and sustainably go it alone, with minimum attention to the health and welfare of whichever part of Earth they occupy—beseeching heavenly help when the going gets tough. Paraphrasing the American historian Eugen Weber, humanity is loth to believe it is not God's special pet. Confused and misled by the premise that only people matter, that our one species is the pinnacle and centre of Creation, people have again and again failed to act toward the Earth in sustainable ways. Ecologically ignorant, we have thought it possible to sustain societies, cultures, economies, without taking as first priority the sustaining of Earth's ecosystems that provide the necessities of human welfare.

Perhaps the realization is dawning that unless Earth and its ecological systems are sustained, human societies can never be sustainable. Homocentric memes, that for thousands of years have taught us our importance, now must take second place to ecological or ecocentric memes that teach the reality of Earth's prior importance.

Now we know that this Earth planet, in whose skin we live, is an immense, vital, integrated system, the Ecosphere. Nothing that we can see, feel, hear, smell, or taste is separate. Everything has co-developed in complex interaction with the rest. The sense of wonder and affection felt for the splendor and bounty of the Earth is natural to us, an expression of our co-evolution with all that exists on the planet. As Rachel Carson opined, it is important *to know* you are part of it, to get out in it, breathe that great air, tramp your feet on the grassy ground, look at the beauty around you, but especially *to get the feeling* that you are a part of it.

We do not understand, and may never fully understand, this miraculous Earth that we briefly occupy. We have made some progress in understanding the intelligence of individual organisms, so that, for example, we know how to plant trees and help them to grow. But we have no comprehension of the wisdom expressed in the structure and function of Earth and its geographic ecosystems. For example, we do not understand the complexity of forest ecosystems that humans have blithely destroyed since the time of Gilgamesh, and continue to destroy today.

Thinking Like an Ecosystem

A long-time ecosystem-thinker and college friend, Dr. Arnold Schultz, lately of the Forestry Faculty of the University of California at Berkeley, proposed that if we want to get along with this Earth-home and adapt our behaviour to it, a good start would be trying "to think like an Ecosystem." What can we learn from the natural ecosystems surrounding us, that is, in addition to silence, beauty, wholeness (the root of holiness), courtesy, and wildness? He noted that ecosystems teach connectedness; they maintain an internal balance, a rough and fluctuating equality between their parts. The health of the whole requires that no one component grows immoderately at the expense of the others. Thus ecosystems teach dynamic balancing of plants and soils and water, a kind of steadiness within limits of plants and animals, between animal populations and available food and water, between production and consumption. In natural ecosystems, this vital equilibrium between parts is maintained by a variety of strategies, such as the negative feed-back control of rabbit populations by predators. Humans, without predators since the dire-wolf passed away, need educational and legal controls to provide the necessary negative feedback, else the dynamic balance between people and the other components of their enveloping ecosystems will sooner or later be upset. Ecological governance is necessary.

But the main point in "thinking like an ecosystem" is to learn *sustainability*, for that apparently is a goal of the Ecosphere and of all its continental and oceanic ecosystems. When left to themselves, ecosystems just keep rolling along, self-repairing and *diversifying*, as they have for millions of years. They only need "management" after we have set other-than-sustainability goals for them, such as maximum timber yield or maximum water yield, after we have "improved" their productivity to serve only human purposes, after we have changed them from complex systems to one-purpose simpler systems. Ecosystems naturally diversify, and so the fostering of diversity rather than simplicity seems a worthy goal for forestry, as it is for agriculture and all other kinds of human culture.

Plantation Agriculture or Forestry?

The word "forestry" is a source of confusion because the one word is applied to two kinds of land treatment—one simplifying and one diversifying—that ought to be clearly separated. An agronomic example shows why. Consider the two ways of "managing" prairie grasslands. One way—called "till-agriculture"—means ploughing down the complex native grassland and replacing it with simple monoculture plantations of cereals, pulse, or oilseed crops. In till-agriculture, after destruction of the grassland ecosystem, crop yield is maintained by subsidies of fossil fuels, fertilizers, biocides, irrigation. It is unsustainable without massive inputs of non-renewable energy. The contrast is "range management," which means preserving the native grasslands, with their hundreds of species of plants and animals in air and soil, while carefully cropping the herbage for hay or pasture. Range management is a sun-run sustainable system that, with care, will go on forever, without subsidies, because the grassland ecosystem is treated as "natural capital," and only a safe percentage of its yearly production, the "natural interest," is cropped. Those who know Merve Wilkinson's woodlot on Vancouver Island and his methods of management will recognize that he practices the equivalent of "range management," and this technique, requiring care and intelligence, surely merits the name FORESTRY in capital letters.

The equivalent of "till-agriculture," and the contrast with true FORESTRY, is "plantation management." Its goal is industrial wood production: clear-cut, scarify, plant genetically selected fast-growing monocultures, destroy the herb-shrub-tree competition with herbicides, pour on nitrogen fertilizer for fast growth, kill insect pests with biocides, grow spaced-and-pruned trees like corn in a corn-field, and when they are hydro-pole-sized cut them down and repeat. In the same way that the average prairie farmer drains the marshes on his land and cuts down the aspen groves, aiming to get every last acre into wheat production, the plantation manager's goal is getting the forestland into wood production. Water, landscape integrity, biodiversity, wildlife, aesthetic and recreational values are marginalized, not because of anyone's evil intent but because (backed by societal ignorance) they contribute little to the bottom line.

Plantation management invites mechanization and the replacement of workers in the woods with big machines. Clear-cutting prepares the stage for plantations, and old-growth forests are the first target. To the plantation manager, the very existence of old-growth forests is a waste. The names "overmature" and "decadent" indicate the attitude. The first priority of the agronomically inclined manager is to log off the slow-growing old forests and replace them with fast-growing plantations.

When I was with the Canadian Forest Service, the professional entomologists and pathologists told me happily that they would never be out of work as long as a main goal of silviculture was planting trees, rather than helping forests to reproduce from seed in their own natural way. Plantations are simplified banquet tables for bugs and fungi, they are recipes for future trouble and hence need plenty of management. From this perspective, every tree planted on forestland is an admission of ignorance and failure. The more we redesign and simplify natural ecosystems on the agriculture model, the more management they need, which of course fits right in with the human proclivity for tinkering. But to foster sustainability, limits to human tinkering must be set for every part of Earth. Two forms of the precautionary principle: "If it ain't broke don't fix it," and "Don't mess around with what you don't understand."

The Forest Wild

Zoning that rules out human uses for large areas is necessary to protect large landscape areas, such as the province of British Columbia, from land-use tinkering. Wilderness preservation is a must. A good start in the interior mountains would be zoning, for complete protection, all the subalpine forests; i.e. the snowy, slow growing, hard-to-regenerate belt of alpine fir/Engelmann spruce, up where the air smells so fine above the cedar/hemlock fir zone. Then protect what fragments of old-growth and ancient forest remain in the lower zones, for these—usually on or close to water-producing sites—are irreplaceable. Foresters in the Pacific Northwest understand that ancient forests are non-renewable resources; once gone they can never be reconstituted. The idea that "old-growth recruitment areas" will ever substitute for logged-off old-growth stands is naive in the extreme, or perhaps more frankly, a devious method of misleading the gullible public.

Several years ago I was astounded to hear a professor, from a prestigious university on the West Coast, telling the world by radio that old-growth forests of the interior mountains needed to be managed by logging, as a sort of sanitation job before fire or insects moved in and devastated them. I wondered how on Earth such forests had got along without management between the time the cordilleran ice-cap receded ten thousand years ago, and the 1890s when prospectors and miners first came into these valleys. How did the magnificent forests survive without forestry's geriatric doctors to anticipate their fatal illnesses, and to prescribe radical surgery as the better way to die? But then I was reminded of a saying by H.L. Mencken, editor and critic, a warning for those who take university education too seriously, "There is no idea," he said, "no matter how stupid, but you can find a professor to support it!"

This ex-professor hopes that his ideas do not fall within Mencken's range. Though not pessimistically inclined, I do from time to time entertain a certain sadness from knowledge of human behaviour in the past, and an apprehension of stupidity ahead. Gilgamesh, the clear-cutter, continues to be society's guiding hero, when attention should more properly be on Ishtar for protection of the forest, and on Mary for protection of the sea.

Hope for better ways of living remains a powerful motivation for change. And so the words of the poet, Pablo Neruda, a lover of the ancient forests of Chile, are appropriate for those who would preserve the beauty of natural forests and of clear, running water:

> I greet you, Hope
> You who come from afar,

You flood with your song the sad hearts.
You give new wings to old dreams . . . [3]

Now is the time to give "new wings to old dreams."

References

[1] Binkley, Clark S., *The Living Dance: Policy and Practices for Biodiversity in Managed Forests* eds. Fred L. Bunnell & Jacklyn F. Johnson (Vancouver: University of British Columbia Press, 1998) Preface viii.

[2] Ralston Saul, John, *The Doubter's Companion: A Dictionary of Aggressive Common Sense* (Toronto and New York: Penguin Books, 1995) 242-243.

[3] Neruda, Pablo, *El río invisible. Poesía y prosa de juventud* (Barcelona: Barcelona Press, 1980) quoted by Lee Yannielli "Earth, People and Poetry: The Forests of Pablo Neruda" *The Trumpeter* (Victoria: LightStar Press, 1997) Vol. 14 No. 1, 1997.

3. THE ECOLOGY OF CITIES

THE ECOLOGY OF CITIES

Reprinted with permission from *The Structurist*, No. 39/40, 1999–2000. University of Saskatchewan, Saskatoon, pp 17–24. Edited for this collection.

Cities are the rich nodes of civilization, the centres of every nation's culture, its commerce, arts, and sciences, which explains why so much attention is focused on city forms, their structures, and their internal functions. Much less attention has been paid to outer ties, relating the city ecologically to its larger geographic setting—the primary focus of this article.

Like coral reefs, cities are complex ecosystems: three-dimensional physical bodies, a close fusion of organic and inorganic components. Analogous to individual organisms, each volumetric city ecosystem depends not only on internal exchanges but also on outside exchanges, relying on the latter for the provision of necessary energy/materials and for the disposal of unnecessary wastes. The far-reaching effects of energy/material inputs and outputs constitute the ecology of these peculiar human-dominated ecosystems.

Confusion results when the inner functionings of cities, their physiology, is mislabelled their ecology. An example is the book *The Ecological City,* a collection of essays that largely deals with internal improvements of urban settlements by designing into them more of the undomesticated world.[1] True, an inner ecology does exist in every urban setting, but it is not the ecology of the city—it is the ecology of people, the connections between inhabitants and the city ecosystem that envelops them.

At the ecology-of-cities level, within Earth's regions, problems are much less tractable than at the ecology-of-people level, within cities. Uncritically mixing the two dissimilar levels fosters an unwarranted optimism about solving city problems. Babylon and Tikal are reminders that city planning and city beautification are no hedge against the dangers of peripheral influences, especially those rendered virulent by neglect.

Civilization = Cities

Cities sprouted from the seeds of agriculture. The tending of crops imposed a sedentary lifestyle, a staying in place to protect the growing food plants and to reap the harvest. The greatest change in human culture after the taming of fire around 200,000 BCE began a mere ten millennia ago when nomadic foragers, enlightened in the art of gardening by women, settled down and turned inventive. Using such technologies as hoes, wheels, and irrigation in fertile valleys they were soon producing surpluses of storable and transportable food. Invigorated by trade and by centralized religious ceremonies, agriculture-based hamlets grew into towns and then into cities. Civilization flourished. Strongmen, priests and kings, took charge. The rest, as they say, is history, which, as we know is always slanted by its tellers who are, of course, city dwellers.

The story of civilization is the story of cities, as shown by the etymology of the two related words. To civilize means to citify, to step out of a primitive or savage condition into a higher existence. To live in the city is to come up in the world, to be cultured and mannerly rather than rude and barbaric, presumably a step closer to Heaven: the "City of God" according to the psalmist (Psalm 46:4).

History slights unlettered rubes and heathens out in the heather. They write no stories and their oral narratives are ephemeral. Histories are the records of civil citizens living in cities, the latter term derived from the Indo-European base "kei" meaning "to camp." Home is the city. "Every Golden Age is an Urban Age," wrote Sir Peter Hall.[2]

Historians recount the wavelike rise and fall of city-based civilizations that gather in a wealth of resources including human talent, develop to a crest, and then decline. In the fallen state reports on the conditions of life are few and uniformly bad. With the glitter of cities gone, Hobbesian times are assumed to prevail—life is nasty, brutish, violent, and short. No news is bad news until, favoured by ameliorating climate, new technology, discovery of unexploited lands, or expropriations through warfare, another round of civilization begins. Revivified through favourable ecological connections, old cities prosper again and new ones appear. Scribes resume the chronicle of important city-based enterprises.

Cities are Not Self-Sustainable

Scrutiny of the rise and fall of civilizations from an ecological perspective reverses the picture. The up-city/down-city phases of civilization's cycle come into focus as "the unsustainable city phase" and "the sustainable non-city phase." The "rise" of civilization, when cities grow, is a transient stage of exploitive living. The "fall" of civilization, when cities collapse, marks return to sustainable living on the land unless, as in the Easter Islands, the limited land base has been ruined.

In the past, cities fell apart when they had exhausted their hinterlands or invaders had done so. Then the population, much reduced in numbers, reverted to foraging and simple subsistence agriculture: the dependable, time-tested way of living by which our ancestors survived.[3] During the early Middle Ages (predictably tagged the "Dark Ages") the treasures of western culture survived in monasteries: forms of rural, horticultural, communal, subsistence living that are models of sustainability and, except for their sexual aberrance, patterns of high culture.

Today cities thrive world-wide, borne to new heights on gushers of oil that make them potential hostages of feudal families and Islamic militants in the Middle East. Hegemonic Western civilization is on a roll, setting the trend of rural depopulation and urbanization for all the world to follow. Fifty per cent of Earth's 6.5 billion people now live in cities and the numbers keep rising. Apart from tangles of vehicles that poison the air, and wretched living conditions for the poor, no insurmountable problems have so far appeared. Cheap transportation based on cheap hydrocarbons has enlarged the hinterlands of cities to include the entire planetary surface. Global trade is ascendant, and all necessities plus a plethora of manufactured instruments of comfort and delight, can be drawn from any part of the globe.

Cities continue to expand and to increase in number while non-human nature degenerates and shrinks. An unhealthy world seems to be the price of a "healthy" (ever-growing) economy whose focal points, according to Jane Jacobs, are vigorous cities rather than the nations that comprise them.[4] All developing economic life depends on city economies because, wherever economic life is growing, the process creates cities. Her nightmare, she has said, is city stagnation globally: no vigorous mature cities, no young cities arising. In such a sad scenario, all is lost; humanity is on the dreary road to universal Ethiopia. Take your pick: either volatile, growing, trading cities, or the back-breaking, backwoods poverty of the primitive foraging society.

There must be other choices. Half a century after Rachael Carson's *Silent Spring* and the dawning of the "Age of Ecology," all glowing commendations of cities as engines of economic growth seem slightly perverse. The thesis that cities are the source of wealth is correct only in the sense that today's human wealth is nature's wealth made over by urban know-how. But a world of growing cities, each building up its industrial plant to reduce dependency on imports so as to become itself a competitive exporter, is a recipe for global disaster. Only the naive can be optimistic and confident that city wealth and city technology will solve the environmental problems they create out of sight beyond their borders.

Historically, the supple Neolithic stage between the two male worlds of Hunter and Urban Broker was a viable way of living. Today's city fathers (city mothers being notably under-represented) would do well to explore modern variations of the old sustainable nurturing option again.

Global Influence of Cities

Six hundred years ago, cities were islands in a patchwork of farms and wilderness. Today they dominate the world and their influence is felt everywhere. Thanks to them the lower atmosphere is a smog soup, soils and food are chemicalized, and millions drink bottled water daily. The human species, adaptable and compliant, accepts all this as unexceptional, as the way of the world. If asked to comment on cities and the lifestyles they entrain, most would undoubtedly simply say: "That's progress." Big cities are the norm today and their central status is unquestioned.

Jerry Mander castigated the media for not ferreting out the relationships, the critical connections, between Earth's environmental problems, current economic theories, and the ways people choose to live.[5] Superficially unrelated happenings—such as clogged highways and climatic change, deforestation and loss of biodiversity, marine technology and collapse of cod and salmon stocks—spring from the same common root: faulty ideas about material progress and about the possibility of ever-higher standards of living for a burgeoning world population. Reigning theories assume the excellence of economic growth and foster its tools of science and technology. They encourage the accelerating use of irreplaceable fossil fuels and Earth's other natural assets. They promote greater global trade and transport, cheering on the adoption of lifestyles that are commodity-intensive. City leadership, exemplified by city growth, is an integral part of this transient splurge.

The decisions, the planning, the lifestyles in cities determine what happens to the nation and to Earth. Therefore, the chief "city problem" is ecological, concerning the impacts of cities on the world outside them. Yet these ecological relationships get secondary attention in urban studies. Most of the latter are physiological, focused on inner functioning, on problems of living within the boundaries. Such analyses are the source of articles on "the eco-city," "the livable city," "the green city." They search for ways to make life bearable in the midst of noise, crowds, flashing lights, traffic, smog, asphalt, stone and glass. The focus on cosmetic rearrangements within cities avoids confronting the mayhem that cities inflict on the regions around them and, by summation, on the Earth.

What needs examination is not urban renewal, affordable housing, architectural creativity, green park spaces, car-free malls, bicycle paths, safe lighting, or even the larger social injustice issues that work against these narrow though highly commendable objectives. The big question concerns the implications of increasing urbanization for the future of the countryside, for global soils and air, for natural areas and wilderness preserves, for lakes and free-running rivers, for forested landscapes patch-skinned at an accelerating rate, for the thirty million other species of organisms to which we are companions. In short, what is the relationship of cities to their hinterlands and, in times of dirt-cheap fossil fuel that allows irresponsible global trade, to the whole Earth?

Cities and their Hinterlands

Lewis Mumford was one of the first to ask ecological questions about cities and their sustainability. His 1965 article, "The Natural History of Urbanization," is precisely on the ecology of urbanization.[6] He noted the dependencies of cities on their hinterlands, and the steady extension of their influences aided by the growing power of transportation technology. He pointed out that the city story, from Nineveh to New York, is one of increasingly substituting the artificial for the natural. Technology rearranges environments so that nature is rarely experienced directly. In losing connections with Earth within the city—those tenuous relationships sustained by gardening excepted—inhabitants also lose track of Earth-relationships without. An illusion of complete independence from nature is fostered, and the phrase "urban sustainability" is no longer recognized as oxymoronic. Mumford expressed pessimism about modern large cities because of their dependence on plunder of the non-urban hinterland.

The same dark thread runs through his *City in History: Its Origins, its Transformations, and its Prospects*, published five years later when cities of the world were under threat of nuclear annihilation.[7] But although Mumford's reading of history repeatedly revealed Metropolis ending in Necropolis, the City of the Dead, he was optimistic that it need not be so. His book is one long sermon exhorting humanity to be better and to do better, not by renouncing the city as a structural/functional error but by converting it, somehow, into what it is not: an environment for fostering community and advancement to a higher-than-economic life, an environment for "man devoting himself to the development of his own deepest humanity," a true vehicle of human progress. This recurring theme expresses his frustrated longing for the Cultural City—a container-museum of mankind's superlative achievements.

The stage for the city was set, according to Mumford, by the first progressive step from man-the-hunter and his clan to the village-based matrifocal agricultural society. The change from wandering forager to sedentary farmer marked "a sexual revolution, a change that gave prominence not to the hunting male but to the more passive female; home and mother are written over every phase of neolithic agriculture." The next quick step upward was from female village to male city.

Mumford's thoughts on the city are ambivalent. He is convinced it is the highest and best structure within which men, drawn by "social and religious impulses," come together "for a more valuable and meaningful kind of life." At the same time he identifies the city as a centre of power, aggression, and aggrandizement, contemptuous of organic processes, contributing to violence and war, eventually inviting its own destruction. Post-city, the people return to rural living-on-the-land in small communities. Then, in the repeating cycle, mother-village nurtures another crop of competitive city-sons.

Each fresh metropolis is reconstituted both physically and ethically by drawing on "the positive forces of cooperation and sentimental communion kept alive in the village." New life, fresh and unsophisticated, "full of crude muscular strength, sexual vitality, procreative zeal, and animal faith" is recruited from rural regions to build new cities or rebuild the old. In Jeffersonian tones Mumford warned that the ancient factor of safety for cities will vanish with the disappearance of the rural source of strength. He accepted the latter as essential for provisioning the city, but not as itself an acceptable environment. To him the rural source is passively female, described as an "unfertilized ovum" needing male insemination to become the urban "developing embryo."

Mumford's writing is riddled with patriarchal assumptions. His protests to the contrary, there is a feeling that the aggressive pursuit of wealth and status in hard-edged canyons of steel is far more valuable than the softer callings of gardening and domestic muddle in the village. From a feminist perspective (if that privilege is allowed me) city culture is a slide back to man-the-hunter, rather than a step up from the matrifocal agricultural society. Significantly, girls like horses while boys prefer motorcycles. Mumford is one of the boys, riding the city-machine.

Cities and the Good Life

What is the city? Originally, and still in Mumford's words, "an instrument for regimenting men, for mastering nature, and for directing the community to the service of the gods," which today are commerce, economic growth, and global free trade. The city is home to *Homo economicus* whose deities are greed and competition. City gods, projections of male fantasies, have always been hegemonic, wrathful, jealous of other gods, and especially resentful of goddesses. Men make themselves in the images of their gods, and they suffer for it.

When celebrants sing praises of cities as places that magnify and enrich human potentiality, that encourage the fullest expression of human capacities and potentialities, that develop the depths of humanity, that store and pass on progressive human culture, one hears an echo of platitudinous, self-congratulatory convocation addresses in praise of the university, itself a city institution. Having convinced themselves of the glories of life in cities compared to the idiocy of rural life, the urbane are blind to the competitive, individualistic dominance theme that works against community. They prefer the hard-boiled eggs of the city to the "unfertilized ova" of the village, and foresee no future other than an urban future.

To the urban intelligentsia, the flaws of the city are surface blemishes that can be patched or glossed over. Assured that the city is the best machine for living, they trust that this time around it will never stop, that Necropolis is forever banished. Sir Peter Hall, the latest city exalter, expresses unbounded confidence:

> No one kind of city, nor any one size of city, has a monopoly on creativity or the good life; but . . . the biggest and most cosmopolitan cities, for all their evident disadvantages and obvious problems, have throughout history been the places that ignited the sacred flame of the human intelligence and the human imagination. Spengler was wrong, for after the sunset comes the dawn; unlike Spengler (and unlike Mumford) this is no tale of decline or disintegration. At the end of the 20th century—eighty years after Spengler foretold the decline of the West, sixty years after Mumford saw the modern city proceeding inexorably to Necropolis—neither western civilization nor the western city shows any sign of decay. On the contrary, this book will be a celebration of the continued vitality, the continual rebirth of creativity in the world's

great cities, as the light wanes in one, it waxes in another; the whole process, it seems, has no end that we know of, or can foresee. The central question, now, is precisely how and why city life renews itself; exactly what is the nature of the creative spark that rekindles the urban fires.[8]

Hall examines neither city ecology nor the dependability of the sumptuous supply of oil and gas that currently stokes the urban fires as it has done since Spengler's time. He writes as one of C.P. Snow's literary intellectuals who hold that culture is autonomous, needing for its ignition no sacred flame nor creative spark from Earth's energetic sources.[9] He puts Mumford down as "fundamentally a brilliant polemical journalist, not a scholar."

I defend the journalist against the scholar. Mumford did acknowledge, at least from time to time, the dependence of the city on what lies outside it. In contrast, the scholar's weighty book—1168 pages with no index entries on ecology, environment, fossil fuels, or energy—assumes no limits to growth and vast possibilities for new Golden Ages of production and consumption, with technology leading the way. Unconnected to Earth, the book is vivid proof that the proper habitat of the city-shaped intellect is fantasyland, aka virtual-reality.

Cities and Virtual Reality

Consideration of the ecology of cities—their relationships to their regions and to Earth—reveals a precarious dependence of which the inhabitants are unaware. Cocooned away from the real world, most city folk live in ignorance of their life-support system.

The "culture" for which cities are acclaimed is itself a kind of virtual reality, providing various compensations for separation from the natural world. Artifice is used to plug the holes in urban lives with facsimiles of important experiences thoughtlessly blown away. The arts serve as humanistic buoys to keep spirits afloat after natural life-preservers have been discarded. Jean-Jaques Rousseau's sour comment that "big cities need plays and corrupt people need novels" can be more positively phrased: "Artists provide food of variable quality for starving souls in cities."

Were Rousseau with us today he would note that, in addition to plays and novels, city people also need the movies, sports arenas, churches, galleries, theatres, zoos, museums and exhibition halls for the display of natural, historical and scientific objects.

In various ways these cultural institutions provide laundered glimpses of the real world of stars, of living landscapes, and organic things. They also offer sanctuaries, escape into quietness or at least, in the concert hall, harmonic relief from simian chattering and city noise. Visit the gothic cathedral to experience the virtual reality of the forest grove and of rolling thunder from the pipe organ. See in the zoo the world's animal inhabitants, and in the planetarium the night sky. Look at the stuffed birds in the museum and watch the nature programs on TV. Drive to the gymnasium and exercise those little-used leg muscles on a treadmill. Immerse yourself in the Olympic-size swimming pool and breathe its chlorinated effluvium, a high-tech substitute for the lake and its bracing air. Study the man-made architecture of old buildings and neglect the marvellous architecture of the world. In short, trust human ingenuity to invent substitutes for nature where it has effectively been walled out. When that falls short of expectation, try Prozak or meditation: therapy for the urbane.

The Ecological Footprint of Cities

In cities the massive import of materials providing food, clothing, shelter, energy and industrial needs, mostly drawn from outside the metropolitan boundaries, is taken for granted. This basic dependency on the wider world is seldom discussed. Veiled also is the equivalent outflow of waste materials and heat that the hapless hinterland must absorb in exchange for its gifts.

A University of British Columbia task force on healthy and sustainable communities developed an ecological accounting tool: the land area needed by average citizens both to provide their resources and to assimilate their wastes at current standards of living.[10] The total "ecological footprint" of each Canadian works out to about 4.2 hectares (ten acres). The ecologist Eugene Odum made a similar calculation for Georgia, a state considered to be sparsely populated.[11] There, in a better climate, the "ecological footprint" of each inhabitant is less than in Canada: about two hectares (five acres). Dividing this into the total area of the state showed it pressing close to overpopulation. Extrapolating studies such as these to the year 2050, several more planet Earths will be required to provide a North American standard of living for the expected 10 billion people!

The impossibility of maintaining today's consumer society in the world of the future suggests curbing the industrial activities of cities and their profligate use of nature's wealth. Economists deem the idea absurd. The market, tuned to inventory, reports abundance; it does not signal unsustainable depletion rates. Further, science and technology will surely come to the rescue should crises loom. Therefore such books as *Reinventing Cities for People and the Planet*, imply that cities will carry on indefinitely as before, made greener by improved management of water, waste, food, landuse, transportation, and energy.[12] The last-named is the keystone that supports all the others. Will the energy supply energize cities forever?

City Ecosystems Run on Energy

The city ecosystem is a sub-system of Earth that operates according to the same rules as a tract of tropical rain forest or a farm; all run on energy, the true currency of the universe.

A small part of the usable energy at Earth's surface comes from its radioactive interior as geothermal energy, but most is provided by the sun in two forms: diffuse daily radiation that seasonally causes trees, grasses, and algae to grow, and the concentrated fossil sunlight stored in plants and animals dead-and-buried long ago, dug out from underground in the form of hydrocarbons that allow cities to function and grow. Industrial civilization is fired up on fossil fuels.

Leaders of governments and their economist advisers are seldom versed in the second law of thermodynamics and its axiom that *energy cannot be recycled*. With each transformation in use, energy is irretrievably lost until, dissipated as heat, it is radiated off into space. Direct solar radiation is renewed every day when the sun rises, but the stored solar radiation in fossil fuels is non-renewable; once used it is gone forever.

As the rich, easy-to-mine hydrocarbons are marketed and burned away, currently at the rate of seventy-five million barrels a day, more and more energy must be expended to find and develop the poorer, more difficult-to-reach sources.[13] Sometime in the twenty-first century the energy required to extract hydrocarbons from Earth will equal or exceed the energy content of what is recovered. Then the net energy—the energy remaining after the energetic costs of drilling, pumping, mining, refining, and transporting have been paid—will be zero. No substitute stands ready to take the place of the power-packed hydrocarbons. The best prospect, nuclear energy, is proving environmentally disastrous. Therefore, economists and their faith in substitutability to the contrary, all human institutions dependent on fossil fuels will literally "run out of gas" and collapse—if pollution or some other catastrophe has not forced the issue earlier. This most realistic of the "doom models" is the least discussed.[14]

The timing of the demise of the world's cities, from Metropolis to Necropolis, hinges on the rate of depletion of fossil sunlight: the non-renewable hydrocarbons. When oil and gas from underground run out, big cities will self-destruct. As supply declines, the smaller cities will be best positioned to survive by decentralizing into sustainable nodes (back to the agricultural village plan, the small garden city). Such a happy ending assumes that, with foresight, the last inventory of non-renewable energy will have been used to establish renewable energy systems: geothermal, hydroelectic, wind, biomass, and direct sunlight. It ought to be happening now.

Cities and Culture

Are cities necessary for healthy psychological and social living? It seems unlikely, given humanity's long ex-urban evolutionary history. The opposite thesis, that cities breed various kinds of unhealthiness, is much more likely.

According to Roszak, city dwellers are psychotic victims of EDD: Earth Deficiency Disease. His diagnosis emphasizes, in another way, that the ecological environment outside the city is more important than the social environment within. Healthy childhood requires opportunities for outside-the-city nature experiences, especially those revealing Earth in its life-giving, food-giving, care-giving role. Roszak remarked that the way the world (the city-as-environment) currently shapes the minds of its children, especially the boys, lies somewhere close to the roots of humanity's environmental dilemma, for the pathological "war on Nature" finds its focus in the industrial city.[15]

As to the social environment, large cities are generally deemed to be the centres of originality and creativity, feeding the flame of culture. Sponsored by city wealth, cultural pursuits are drawn into the urban machine which is then confirmed as their natural home. Much of what is popularly considered culture in the creative arts and sciences is disconnected from reality by the very fact that the city is its milieu. City culture is refined in a virtual-reality crucible, a container insulated from its roots, out of touch with its source. Such a culture is likely to be blind to Nature's truths, unintentionally contributing to and supporting urban fantasies of which there are many.

The intellectual religions of the world are the mind's supreme virtual realities. City-based and patriarchal, they renounce animism with its vital insight that Mother Earth is the life-source. Technology is the body's supreme virtual reality. City-based, it renounces Nature and substitutes artifice for the essential processes of Earth. "Virtual mind/body" is perfectly realized in the city and with it "virtual culture" that has little connection with Earth.

Conclusion

Suppose that the end of cities as we know them can be foreseen as the time—twenty, thirty or forty years from now—when the flow of irreplaceable, non-renewable resources that support civilization slows to a trickle. To survive and thrive, humanity will have to live a different kind of life based on different values and different attitudes. City culture, the culture of civilization, will be passé, not because it served humanity badly but because on a depleted, impoverished planet it is irrelevant. City culture will be a museum piece remembered for its wanton ways, displayed in another kind of Earth environment.

In tracing the history of deforestation for city-building through the ages, Perlin commented that "civilization has never recognized limits to its needs."[16] Nothing today contradicts that statement. Its truth means that cities and city culture cannot be beacons for the future. The source of a sustained human culture will always lie outside the city, in closer ecological relationships to Earth.

The twentieth century neglected the ecology of cities, assuming that they will go on forever. The imagining, designing, and testing of sustainable alternatives to cities and city culture by the arts and sciences is the task of the twenty-first.

References

[1] Platt, Rutherford H., Rowan A. Rowntree, and Pamela C. Muick, eds. *The Ecological City* (Amherst: The University of Massachusetts Press, 1994).

[2] Sir Peter Hall, *Cities In Civilization* (New York: Pantheon Books, 1998).

[3] Quinn, Daniel, *The Story Of B.* (New York and Toronto: Bantam Books, 1996).

[4] Jacobs, Jane, *Cities and the Wealth of Nations* (New York: Random House, 1984).

[5] Mander, Jerry, and Edward Goldsmith, eds. *The Case Against the Global Economy, and For a Turn Toward the Local* (San Francisco: Sierra Club Books, 1996).

[6] Mumford, Lewis, "The Natural History of Urbanization," *Man's Role in Changing the Face of the Earth* ed. W.L. Thomas Jr. (Chicago: University of Chicago Press, 1956).

[7] Mumford, Lewis, *The City in History: its Origins, its Transformations, and its Prospects* (New York: Harcourt, Brace & World, Inc., 1961).

[8] Hall, Ibid.

[9] Snow, C.P., "The Two Cultures," *The New Statesman*, 6 October 1956.

[10] Rees, William E., "The Footprints of Consumption: Tracking Ecospheric Decline," *The Trumpeter* Vol.14 No.1 (Victoria: LightStar Press, 1985) 2-4.

[11] Odum, Eugene, *Ecological Vignettes: Ecological Approaches to Dealing with Human Predicaments* (Amsterdam: Harwood Academic Publishers, 1998).

[12] O'Meara, Molly, *Reinventing Cities for People and the Planet* (Washington, DC: Worldwatch Institute Paper 147, June 1999).

[13] Hanson, Jay, "Energetic Limits to Growth." *ENERGY Magazine*, Spring 1999. http://www.dieoff.com/page175.htm

[14] An editorial, "Models of Doom," in *The Economist* (20 December 1997) suggested that Malthus in 1798 was the first pessimistic environmentalist to get it all wrong, and that his successors have been as mistaken as he. Perhaps, but no one until recently has looked at the net energy equation on which city civilization depends.

[15] Roszak, Theodore, *The Voice of the Earth* (New York & Toronto: Simon & Schuster, 1992).

[16] Perlin, John, *A Forest Journey* (Cambridge MA: Harvard University Press. 1991).

ECOLOGY AND ARCHITECTURE

Reprinted with permission from *The Structurist*, No. 31/32, 1991–1992. University of Saskatchewan, Saskatoon, pp 16–22. Edited for this collection.

> "*We make our environments and then they make us.*"
> —Winston Churchill, arguing for full restoration of the bombed Houses of Parliament.

> "*Big buildings; small people.*" —Stephen Leacock

The vital context of people, and their experiences, is a surrounding interpenetrating world: Nature, Earth, the Ecosphere. Brought into being by the blue planet, immersed in its living skin, clustered where its gas/liquid/solid phases meet, people commune constantly with their host—breathing its air, drinking its water, eating its soil through the bodies of plants and animals, sustained by its energy, attuned to its diurnal and seasonal rhythms, inspired and soothed by its beauty, awestruck in moments of insight by its improbable unity and miraculous integration.

The marvel is that such a cohesive, self-balancing creation should produce within itself a species with a consciousness that seems to be alone and autonomous. The featherless biped is cursed with an ego, a sense of separate selfhood that flies in the face of ecological reality. The child, we are told, *naturally* rejects its mother in order to declare an independent selfness, which is thereafter encouraged—under the rubric "growing up"—by the family, by the school system, by the culture. Praise of skills, talents and possessions feeds selfness and selfishness whose accretion is soon self-sustaining, fostered by social conventions, by strivings for personal success and recognition, by competitive games and competitive business.

Applauded by compatriots, the actor emerges from the supportive wings of the Earth-theatre, advances to centre stage, recites the lines of personal authenticity, and declares *this self* to be the prime reality, the measure of all things—a conclusion at odds with her/his beginnings and endings.

From solipsism, from the sense of "only me," the arts and sciences take their cue, for human endeavors in their various forms embody prevalent ideas, concepts and values. In architecture, for example, the International Style of the 1920s made a virtue of stand-aloneness, its symbol the phallic skyscraper advertising the power of its owners. Architects bemused by technology, unwilling or unable to devise symbiotic structural arrangements, designed glass sheathed, hard-edged, flat-topped Modernist buildings to be dropped anywhere in the world regardless of environment, matched only to motorways serving fleets of private vehicles. Today's Postmodernism and Neomodernism are eclectic but without direction, its practitioners not so much denying their predecessors' philosophy as rejecting its outcome in soulless "machine-precision" towers and boxy buildings. Underlying ideas are untouched. For what else is there in the popular ideology but glorification of the individual and praise for the aggressive standout, the egregious free-market entrepreneur?

In this article I examine the cult of the personal, asking if the needed counterbalance may not be found in a public life based on ecological understanding, an extension beyond the social and communitarian. I argue that we are ecological animals first, and this fact gives point, secondarily, to our sociality. Shaped by the ecosystems in which we have lived and now live, the challenge is to break out of our species-centredness, our homocentrism, and in the arts explore the dictum, "form follows Ecology."

Private Life

The spirit of capitalism, said Weber, is the Protestant ethic understood as "worldly asceticism"—the renunciation of social gratification both from ritual religion (denial of Catholicism) and from immediate spending of one's money (in the interests of accumulating capital).[1] To these twin roots of secularism and capitalism, Sennett traced the erosion of public life and the ascendancy of private life whose outcome is narcissism—the assumption that the psyche of the feeling person has an authentic inner life of its own, independent of any kind of outer "impersonal" life.[2]

Although not identified as such, Sennett's argument against the ascendancy of the "personal" life is essentially ecological: each of us needs consciously to pursue an outside-the-body life, an impersonal life in interaction with the larger-than-personal world, in order to develop a worthy personality. The to-and-fro dialectic of society and the individual, of the impersonal-personal, is essential; an empty public life results in an unbalanced personal life. Put another way, each of us does not make, control and *authenticate* the self, in the words of Taylor,[3] so much as discover, by interaction with the larger world, one's *human nature*.

Without that interaction as an axiom of architectural design, the public domain loses its importance, functioning only to serve the private domain. The appealing "wild" of city streets is being appropriated, forced indoors, tamed under glass, de-politicized, privatized and policed to exclude democratic friction and vitality.[4] The public domain is reduced to thin spatial connections conveying traffic between buildings wherein individuals are separated and cocooned off from their natural environments and from each other. Many are the examples of the results of neglecting, in design, the relationships of *outer space* to *inner space*, eventuating in noisy car-corrupted corridors, wind-swept canyons walled with glass and steel, threatening concrete concourses, and various other kinds of unused, dead public spaces. The more city centres are built upon, the more they lose their original small-town charm and the greater the likelihood of their turning into fearful places.

Inattention to design in the public domain fosters deterioration of the city. Urban citizens, vaguely feeling that something is missing, rationalize the lack as an absence of "community" and attempt to create a spirit of fellowship by extension of those personal and intimate feelings that spring from the private domain, from individual and family. This camaraderie is only possible within small subdivisions of the city, within districts or blocks, and even then such feelings cannot long be sustained.[5] The "community" as an expression of shared emotions is narrow and parochial; it cannot hold together except by fighting city hall or some other group considered invasive or otherwise threatening. The energy that should be turned toward designing a better city is frittered away among city sectors, often in competition, each striving for an imagined intimacy, for a validation of familial emotions. The magnetic pole of the impersonal—historically a compelling force in the public life of the wider world—is missing.

In former times, said Weber, a prophetic *pneuma* or "spirit" swept through great communities like a firebrand, welding them together, but today only a muted residue survives, pulsating *pianissimo* in personal human situations, "The fate of our times is characterized by rationalization and intellectualization and, above all, by the disenchantment of the world. Precisely the ultimate and most sublime values have retreated from public life either into the . . . realm of mystic life or into the brotherliness of direct and personal human relations. It is not accidental that our greatest art is intimate and not monumental."[6]

Philosophic Liberalism

Other sources, political and literary, feed the notion of individuality and militate against balanced concerns for the impersonal and the ecological. Rebellion against kingly power in the Americas and in France during the eighteenth century endowed the Western world with a fervent belief in individualism, a faith in the rights of each person to liberty and to possession of property as protection for that liberty. Both the American and the French constitutions were made for the individual and represent philosophic liberalism, elevating the importance of the person relative to the social context and to the worldly milieu. It has little to say on the subject of compensating responsibilities, about rights for the other-than-individual. Still, today, the two-hundred-year-old theme is repeated verbatim. Fearful of the State that has usurped former royal powers, some are demanding the entrenchment of property rights in charters and constitutions, hopeful that the codified law will prevent future curtailment of the freedoms of individuals to own and remake Earth's properties as they please.

In an article titled "Killing Wilderness," Wayland Drew compared the anti-utopian books, *We* by Zamiatin, *Brave New World* by Huxley and *1984* by Orwell.[7] He noted their shared libertarian tradition. All three warn that the rapidly developing managerial-technological society is to be feared, no matter what political system accompanies its development. A machine-perfect society will necessarily be totalitarian because its goal is *efficiency*, while the free autonomous individual—skeptical, irrational, passionate, ecstatic, raging, agonizing, heroic, honorable and therefore recalcitrant in the extreme—is by nature *inefficient* in the same sense as this diverse unpredictable Earth is inefficient. Drew concluded that the best means of stalling or derailing the technologic juggernaut is by championing the concept of inviolate wilderness, saving from human interference large undomesticated parts of Earth.

With this conclusion I agree. Major wilderness preserves seriously threaten the advance of technology. By their very existence, wilderness areas challenge and potentially subvert civilization's goals of control and management. But is the purpose and justification for establishing and guarding wildernesses only to salvage the freedom of human individuals, as Zamiatin, Huxley and Orwell suggest? Drew hinted at another justification when he characterized wilderness as "the matrix of life itself," a theme developed in his novel *The Halfway Man*.[8] Wild existence is its own reason for being. Wilderness displays "nature creative," the source of all animation, inherently worthy, beautiful and "right," far beyond its many values for the people that some millions of years ago it conceived and brought forth. The wild represents the missing *impersonal* pole, the neglected ecological pole, the magnetic concept that, teamed dialectically with the *personal* pole, might again ignite a worthy public life, perhaps a source of the *pneuma* that in former times, in Weber's words, "welded great communities like a firebrand."

Huxley[9] and the other dystopian novelists did not confront the dangers of individualism, of the egoism engendered when the highest cultural good is individual liberty. In a society such as ours, the political trend is toward a repressive oligarchic State, for such is the logical outcome of individual egos seeking political power without the restraint of strong public goals. If the individual is accepted as the most important or only important entity, then those who by grace, good fortune or cunning obtain political power will be strongly tempted to maximize their own freedom and that of their class by dominating the less fortunate. The wealthy in public office tend to ignore the poverty stricken. Rich and powerful corporations, the ultimate in selfish bodies, take over the political and educational systems to keep poor "consumers" in line. The regimes feared by the dystopian novelists, those that deprive their citizens of freedom, are precisely the fruits of liberal philosophy.

The natural expression of the "democratic instinct" seems to be the struggle of the individual against an oppressive State and its corporate allies. Such a struggle is worthy, but it cannot be successful as long as both friend and foe subscribe to the same doctrine of unchecked individual freedom. Idealizing the individual as freedom-seeking hero does nothing to diminish rampant egoism whose inevitable political expression is the oligarchic State in which a relatively few are "freer" than all the rest. The outcome of elevating the freedom of the individual as the greatest good leads to loss of freedom for the majority—unless strong countering values are firmly anchored in some kind of impersonal and public world.

In-turned individuality confirms personal freedom as an inalienable right, and society as only a "social contract," a prudent coming together of individuals in order to protect themselves, their property and their concerns. The chief interests of sociology become the constraints that the community and nature place on the creativity of the individual.[10] If "society" is a myth, no more than a contract invented to secure the interests of autonomous persons, why should they not co-opt it for their own advantage? Assuming that society is designed to serve me, why should I not subvert it if, by that subversion, I and my class are better served?

Ideas of social responsibility as sponsors of justice and equity are inadequate when, relative to the importance of the individual, the importance of society has faded. Social obligation (*noblesse oblige*) commands no "political will," no sense of great importance. Society is seen to consist of winners and losers, and the latter—according to current conservative theory—have nobody but themselves to blame. Democracy may require as its prerequisite a self-reliant people with a strong sense of personal self-esteem[11] but without a collective conscience, wider than personal, it is sure to founder.

Tomorrow's Public World: Ecological Society

The social contract theory of society can be countered by communitarian arguments that point out the priority in time of the social as compared to the individual: the mating of the sexes before the baby, the family as a survival unit, tribal care extended to the young and the infirm, the necessary social context of language and culture. Nevertheless, "society" as an aggregation lacks the substantiality of the "individual." Unlike the person—a solid functional form in space and time—society is a vague taxonomic category, a gathering of people that share the same geographic area and, in general, the same fundamental values and ideas. Society is more abstract than person, just as forest is more abstract than tree.

"Society" gains the same substantiality as "person," and "forest" the same substantiality as "tree," when to each is added its *geographic matrix*; that is, the atmospheric layer over it, the soil/water layer under it, and the companion organisms around it. Then both "society" and "forest" are made whole. Each is converted from an abstract community into a functional ecosystem. Thus conceived, every human society is an organic part of an ecosystem within a particular geographic ecoregion. Frederic Clements, an early North American ecologist, coined the term "ecesis" and defined it as the process whereby organisms get established in place, making themselves at home, making themselves partners with air, soil, water and other organisms.[12] Humanity needs to attend more to "ecesity," to making a home harmoniously within the Earthly milieu. Here is new meaning for the sociological abstraction, "society," and a signpost for a more rigorous "sociology."

Ecesis posits no self-serving contract between individuals but rather conceives them as parts related within a greater ecological whole in whose healthful functioning they participate. Political order in the service of society may be only a matter of agreement among individuals; ecological order follows laws of a higher level, not those enacted by one species for its convenience. Post-Neomodernism or post-humanism should be informed by a comprehension of humanity's ecological status, enlarged beyond the navel-gazing that has so far characterized the history of our species.

We should be asking how the things we construct—the highways, buildings, aggregates of buildings, towns and cities—connect us to the enveloping Ecosphere. Do they help or hinder making connections, are they opaque to environment or lucid, do they enhance sensory perceptions of the wider world? Are they vital, life-enhancing, mythic and poetic? Or boring, stultifying, colourless and disharmonic within and without? Are they "organic" in the sense championed by Frank Lloyd Wright: do they love the ground on which they stand? When creative architecture answers yes, said Wright, it is the centre line of all culture.[13]

The Arts and Ecesity

According to Eli Bornstein, "One of the frontiers of our time—a frontier often neglected—is the organic unification in art and architecture of structure and colour in space and light. Are these elements homologous? Can they combine to flourish symbiotically as a new poetic of art and architecture: poetical buildings, artistic forms, a holistic synthesis of multidimensional elements of painting, sculpture and architecture?"[14] Much depends on the wisdom, on the ethical prescience of people, for in the words of Louis H. Sullivan, "every building you see is the image of a man whom you do not see . . . our buildings, as a whole, are but a huge screen behind which are our people as a whole."[15]

An affirmative answer to Bornstein's question presupposes a shared compelling vision, enlarged far beyond the traditional focus on individualism and, by extension, on human society. Such a vision must have a spiritual dimension, meaning by "spirit" the universal liveliness, *animus* or *pneuma* that informs the world and everything in it, not just the animate but the miscalled "inanimate" as well. Fortunately, a new and inspiring Nature-revelation is today unfolding, one implied a few decades ago when architect Richard Neutra wrote his essays on *Nature Near* [16] and landscape architect Ian McHarg wrote *Design With Nature*.[17]

Planet Earth is an integrated system, born 4.6 billion years ago at the collapse of a supernova, miraculously differentiating and organizing itself ever since. It has evolved, and continues to evolve as a creative entity, a life-giving whole whose components are interrelated and interdependent in many marvelous ways that we are just beginning to understand. Seas, continents, atmosphere and organisms are integral parts of the Ecosphere, changing and evolving synchronously, and none could exist without the others. Thus humans are revealed as one of thirty million compatible species of organisms, all from star dust, evolved within and sustained by the inorganic-organic Earth whole, adapted to its geographic parts and their air/water/soil/organism constituents hitherto understood as "our resources." The great forest belts—placed here, we thought, by divine providence for our consumption—are, among other things, evapotranspiration pumps that affect global precipitation. The soils that we have plundered for agricultural wealth are organs of vegetation reproduction. The seas and the atmosphere, whose chemistry we are busily changing, are temperature ameliorators and material transporters. All Earth's parts are the stuff of life.

Life is a property of the Ecosphere and its sectoral ecosystems, not of stand-alone organisms. The geological strata at Earth's surface, the soils, sediments and water bodies, the air and received sunlight are as much *alive* as organisms, for none can exist without the others. The *animus/anima* informs the whole world, though expressed most vivaciously in things like us. A corollary: we are parts not wholes, totally dependent on the Earth. That parts serve wholes, rather than the reverse, is a truism. The function of humanity is not the witless reproduction of people and machines unendingly, but the protection of the Ecosphere's healthy functioning, ensuring the preservation of its creativity and power. This means constant awareness that we exist within a larger Being.

Form Follows Ecology

Louis H. Sullivan argued that the form of a building, by analogy with organisms, should express its function. The idea is worth exploring in a wider, ecological perspective, considering architectural structures in the context of their local and/or regional ecosystems, defining the latter as geographic places: specific three-dimensional chunks of Earth-space that include societies and their artifacts.

Function has two meanings, indicated by the questions, "How does it function?" and "What is its function?" The first is physiological, concerned with what is going on within some thing of interest. This I take to be Sullivan's focus: the morphology or form of a building should be consistent with its internal uses. The second question is ecological, asking how a thing relates to what envelops it.

The internal/external division, differentiating function-within from function-without, is somewhat arbitrary because the "skin" or membrane that separates the inward from the outward is permeable. Logic suggests that the Earth-surface ecosystem encapsulating each organism is primarily determinative of both external and internal form. Marshes shaped ducks and muskrats, savannas shaped horses and humans. In the organic world, ecological conditions shape forms naturally; in the built world, the creative world has first to be thought. Usually that thought has been physiological.

It was not always so. Cities used to be constructed as images of the cosmos, temples were raised as symbols of psychic wholeness, domes were likenesses of heaven's vault, and forms such as the ziggurat linked Earth and sky.[18] In recent times few architects have made it an article of faith that forms should respond literally or symbolically to surrounding nature, adapted creatively and aesthetically to their regional ecosystems. Frank Lloyd Wright came close with his ideas of an organic architecture, a poetic architecture "growing out of the ground" according to its own appropriate nature, free of the extraneous, in recognition of the fact that by nature we are "ground-loving animals." His was a commendable vision of an organic society wherein art, religion and science provide the unitary context for a compatible architecture.

Imaginative ideas are also present in northern North America where regionalism, combining the vernacular and the landscape, is expressed in recent Canadian architecture. Both Arthur Erickson and Douglas Cardinal have followed ecological visions. Erickson's early architectural work, like that of Wright, was guided by a poet's rhythm and a painter's vision, his richest source of inspiration "the dialogue between a building and its setting." He has claimed to be more landscape designer than architect, and such a mind-set—focused on a landscape approach—necessarily attends to "spirit of place" while attempting to be socially and environmentally conserving.[19]

Cardinal's indigenous architectural style is mythically in dialogue with landforms and their geological histories, with rock outcroppings and streams. Boddy characterized Cardinal's designs as "organic-cum-geological," expressing landscape spiritually transformed:

> The shell that Cardinal designed for the museum (the Canadian Museum of Civilization in Ottawa/Hull) drew its inspiration from the landscape—which is his specialty—but particularly from the landscape of the distant past: the end of the Ice Age when humans first set foot on this continent. One wing of the building communicates an image of the Canadian Shield bedrock, its angular outcroppings smoothed by the overriding glaciers, and then undercut by outwash streams from the melting ice. The other symbolizes the thawing glacier itself—the huge window-wall of the Grand Hall—exposing the eskers and drumlins of gravel and glacial till, gradually re-colonized by vegetation (the great copper vaults, that will slowly turn from brown to green). Between the two wings, plazas and parkland represent the plains over which Canada's first immigrants travelled, millennia ago, as they made their way across the continent.[20]

As these examples show, artistic form in the human world follows ideology. The shape of structures is first imagined according to the artist's beliefs, then brought into being according to ideas that society is willing to support. The more we understand this miraculous world, the more we marvel at its mystery and creative power, the more empathic our feelings for it, the more our art forms will reflect a celebration of it. If in the future homocentrism yields to ecocentrism, if the ideology of species-selfishness gives way to a more generous paradigm, then built-forms will closely follow the promptings of ecology.

Conclusion

Weber traced our malaise to the disenchantment of the world, a loss of that outward bonding without which individuals find their lives less and less rewarding. If the world is to become "re-enchanted," certainly the efforts of artists and the artistic will be the means. What better place to start with than the symbolic sculptures within which we live?

In his opium-induced trances, Thomas De Quincy envisioned magnificent architecture, "such pomp of cities and palaces as never yet was beheld by the waking eye, unless in the clouds . . . cities and temples beyond the art of Phidias and Praxiteles—beyond the splendors of Babylon and Hekatompylos."[21] The same bejewelled revelations, in vast landscapes bathed with preternatural light, have presented themselves to other visionaries. By these Aldous Huxley was prompted to wonder if all humans do not carry subconsciously, at the antipodes of the practical mind, dramatic archetypal designs joined with vistas of surpassing beauty that erupt into consciousness as inspirational forms, whenever the doors of perception are cleansed: when the crass promptings of utilitarian reason are stilled and the purely imaginative is allowed to shine through.[22]

The materials of a building, said Sullivan, are but the elements of Earth removed from the matrix of nature, and reorganized and reshaped by force—by force mechanical, muscular, mental, emotional, moral and spiritual. If these elements are to be robbed of divinity, he said, let them at least become truly human. Perhaps to "become truly human" is to pay more heed in artistic endeavors to the morality, spirituality, divinity of the world from which we have taken so much and to which, so far, we have given so little.

References

[1] Weber, Max, *The Protestant Ethic and the Spirit of Capitalism* (New York: Charles Scribner's Sons, 1958).

[2] Sennett, Richard, *The Fall of Public Man* (New York: Random House, Vintage Books, 1978).

[3] Taylor, Charles, excerpt from "The Sources of Authenticity," *Canadian Forum* Vol. LXX, No.806, January/February 1992, 4-5.

[4] Boddy, Trevor, "Underground and Overhead: Building the Analogous City," *Variations on a Theme Park* ed. Michael Sorkin (New York: Noonday Press, 1992)123-153.

[5] Cox, Harvey, *The Secular City* (New York: The Macmillan Company, 1966) 40.

[6] Weber, Ibid.

[7] Wayland Drew, "Killing Wilderness" *The Trumpeter* Vol. 3 No. 1 (Victoria: LightStar Press, 1986) 19-23.

[8] Drew, Wayland, *The Halfway Man* (Ottawa: Oberon Press, 1989).

[9] Huxley, Aldous, *Brave New World and Brave New World Revisited* (New York: Harper Colophon Books, 1965) 21.

[10] Westhues, Kenneth, *First Sociology* (Toronto: McGraw-Hill Inc. 1982).

[11] Steinem, Gloria, *Revolution from Within: a Book of Self-Esteem* (Toronto: Little, Brown and Company, 1992).

[12] Clements, Frederic E., *Plant Succession* (Washington: The Carnegie Institute, 1916).

[13] Wright, Frank Lloyd, *An Organic Architecture: the Architecture of Democracy* (London: Lund Humphries & Co. Ltd., 1939) 4.

[14] Eli Bornstein, ed. *The Structurist* Vol. 29/30 (Saskatoon: University of Saskatchewan, 1989/90) 132.

[15] Sullivan, Louis H., *Kindergarten Chats and Other Writings* (New York: Dover Publications, 1979).

[16] Neutra, Richard *Nature Near: Late Essays of Richard Neutra* (Santa Barbara: Capra Press, 1989).

[17] McHarg, I. L., *Design with Nature* (New York: Natural History Press/Doubleday, 1969).

[18] Tuan, Yi-Fu, *Topophilia, A Study of Environmental Perceptions, Attitudes and Values* (New Jersey, Englewood Cliffs: Prentice Hall, Inc., 1974)129-138.

[19] Cawker, Ruth, and Wm. Bernstein, *Contemporary Canadian Architecture* (Toronto: Fitzhenry & Whiteside Ltd., 1988).

[20] Boddy, Trevor, "Preface" *The Architecture of Douglas Cardinal with Essays by Douglas Cardinal* (Edmonton: NeWest Press, 1989) 1.

[21] De Quincey, Thomas, *Confessions of an English Opium Eater* (New York and Toronto: The People's Library, Cassell and Company Ltd., MCMV111) 178.

[22] Huxley, Aldous, *The Doors of Perception* and *Heaven and Hell* (New York and Toronto: Perennial Library, Harper & Row Publishers, 1954).

EDUCATION FOR A NEW WORLD VIEW

Can education help in the cultural healing that seems necessary to cure a sick environment? Playing devil's advocate, David Orr suggested that education is not the solution but part of the problem[1]. Knowledge is power, and the more we know the more we throw our weight around. After all, well-educated people, not illiterates, are wrecking the planet. Schools and universities are morally bankrupt, most research is worthless busywork, while new knowledge creates increasing ignorance of how to handle it—witness the exposure of our ignorance of the atmosphere by the CFC ozone-destroying fiasco—and so he asks, "Is Conservation Education an oxymoron?"

Orr's point is that education continues to serve an outworn system that fabricates human-wealth "goods" at the escalating cost of "bads" for the Earth: the extermination of other species, pollution of water, air and land, deforestation, the degradation of soils, expanding deserts. Top-loading the system with ever more data and facts to create the semblance of an "informed society"—a traditional role of education—does not counter the world's increasing instability. Western culture is disoriented not so much from lack of information as from wrong values and beliefs, chief of which is the conviction that humans have the God-given right to dominate, control and manage the entire Earth and all that is in it. If education proceeds from this axiom it is indeed a threat, adding fuel to environmental conflagrations rather than helping to damp their flames.

Fortunately the attitudes that are sapping the environment's health and sabotaging prospects for a viable future are not instinctive but cultural. They can be changed by better understanding the planet's history and our own natural history, by better appreciating the evolutionary story that relates us to other organisms, and by better comprehending ecology that dissolves the boundaries between ourselves and the landscape ecosystems within which we live. Comprehension generates its appropriate values, chief of which—with things revealed to us as important—are sympathy and care.

Should values be taught in school? Inescapably they always are. Whether parents like it or not, ethical standards are implicit in all subjects taught to their children. Recognition and acceptance that education is necessarily normative clears the way for more important questions: What values should be taught? What values are in harmony with ecological truths? Then education's task is clarified as leading in new directions, away from conventional beliefs/values/attitudes that the realities of a worsening world are proving false. The duty of education is to foster a new frame of reference for all that people think and do, a reorientation toward deeper and truer insights with power to set this and succeeding generations on a more charitable and creative path vis-a-vis the surrounding world.

So Little Ecology for the Mind

The historian and educator Hilda Neatby recounted a story about Lincoln Steffins, editor and author, who travelled to Europe just after his graduation from university.[2] The turning point in his education came when he met a group of Oxford men and listened to their discussions. "Those men never mentioned themselves," he said. "Their interest was in the world outside of themselves . . . their conversations . . . established in me the realization that *the world was more interesting than I was*." Neatby added, "Here is as good a definition of education as any: the discovery that the world is more interesting than oneself. It is also a good definition of citizenship, and of mental health."

Hindsight from the 1990s shows the revelation is deeper than Neatby, Steffins and the Oxford men suspected, for in their time the appreciation of human ecology was weak and "the world" meant only the universe of society and culture, comprising human thoughts, words and deeds. The matrix of civilization, the natural world from which the built and cultural environments are drawn, was unimportant. Only people mattered. Works of art drew metaphorically on nature but without a deep sense of its importance. T. S. Eliot's popular poem *The Waste Land*, written in 1922, is less a prescient judgment on the state of the world than a melancholy commentary on a failed society, and on Eliot's failed first marriage.

The literal truth of the statement that "the world (planet Earth and all its contents) is more interesting than we" is gaining converts. Human interest is attracted not only by Earth's beauty but by new knowledge of its integrity and creativity—of which humanity is a recent outcome. The moon-shots of Earth from the outside, showing an integrated ecological system, place our species in correct perspective as the planet's major stress rather than as partner in a cooperative whole.[3]

In revising school curricula, both perspectives, that of the planet-as-whole and of people as potential participants within it, cry out for primary attention. The two together, comprising human ecology, should be the core of education. Without the revolutionary understanding of human ecology, civilization will continue on its self-serving and suicidal way. Without the guidance of ecological comprehension all traditional curriculum subjects are likely to prove useless, posing dangers to sustainable living on the planet.

Appreciating the Planet as Ecosphere

Western religions, arts and sciences reflect a cultural obsession with two relationships, that of Man to God, and of Man to Man. Earth and women—symbolically related and stigmatized as inferior—have largely been ignored. Their joint re-discovery today is the other side of a growing suspicion that what comes naturally to the combative male in an aggressive economy is no longer appropriate. Widespread environmental degradation is the sign that the limits to belligerent exploitation of a fixed Earth-space have already been exceeded. A less masculine approach to the people-Nature relationship is central. Henceforth, a higher degree of care, compliance and cooperation will make for a sustained and better home.

The home to which we belong and to which we must adapt is the Ecosphere, literally the Home-sphere. For 4.6 billion years it has evolved under beneficent sunlight as a unitary being with related parts—atmosphere, hydrosphere, lithospheric sediments, and organisms—in an ongoing process that will continue for several billion more years. Organisms are one kind of differentiated *part*, integral with air, water and soil of which they are composed. Late comers such as we, tracing our history of humanness for only a few million years, nevertheless stem from an ancestral line that goes back to the beginning of time. Our species is one twig on a branching tree, rooted in air/water/soil, whose bacterial origins have been found imprinted in ancient rocks three and one half billion years old. Evidently the inorganic and organic are intimately inter-related.

Life is a function of the Ecosphere and its sectoral land and water ecosystems. Protecting and maintaining *them* is the only way to protect and maintain life, including our lives. To separate Earth's surface into animate and inanimate, biotic and abiotic, living and dead, devalues the inorganic as "mere environment." It sets up organic things like us as all-important, allowing exploitation and ruination of the rest. Education needs the truth that Earth and its geographic places (geoecosystems) are the important entities—the matrix (mother) of the inorganic/organic species they encapsulate. Once this bedrock fact is established, the various disciplines—mathematics, physics, chemistry, biology, sociology, the humanities—will be reoriented in life-supporting directions.

Words are important. The weak word "environment" must be replaced by the strong terms Earth and Ecosphere (literally the Home-sphere). The only serious threat to the survival of the human race is continued destruction and poisoning of the Ecosphere's functional parts, unrecognized as such because we have labeled them "God-given resources," an ignorant rationalization for our careless plunder of seas and soils, grasslands, forests and rivers. We have thought of them all as unfinished things to be "developed"—as if, by analogy, our hearts, lungs and skeletons could be processed and manufactured into improved products for the welfare of one type of body cell.

Concepts are important too. The Home-sphere concept makes ridiculous many conventional attitudes. It provides an antidote to human self-importance and selfishness, to ethnocentrism and to chauvinism. Its acceptance provides hope that, with understanding and compassion, all the world's people might be turned from internecine strife and brought together on a common mission: to curb destructive human actions and thereby serve and save the globe, and with it the best of human culture. This should be the meaning of "global education," providing an integrative framework for important subsidiary subjects such as the cultures of the world's peoples.

People Participants in the Ecosphere

Our ancestors were not ecologists. They did not understand that organisms—themselves and things like them—were parts of a greater whole. The story of philosophy since Plato, and the story of religion in the Judaeo/Christian/Islam traditions, clearly show that only people have mattered. On this questionable base the liberal tradition that grew out of the eighteenth century Age of Enlightenment built a strong case for individualism: the primary reality is the person, an object of intrinsic importance with inalienable rights to life, liberty and the pursuit of property. Society is secondary, only a prudent gathering together by contractual agreements in order to protect the rights of individuals.

In reaction, the liberal view has been challenged by a parallel communitarian tradition claiming that people are better understood as social beings, sexual beings, cultural beings; without the support of others they cannot exist, reproduce or learn to think. Therefore each person is *a part* of the collectivity; the vaunted rights of individuals have no meaning without social responsibilities that provide the foundation for expression of individual differences, for individual creativity.

Around this duality of individual/society, political and economic theories are organized. Will it be liberty or equality, freedom or justice, privatization or communal ownership, the free market or the planned economy? Yet both sides share a common people-centred philosophy that implies an indifference if not antipathy to Nature. On the one side is our self-congratulatory species marvelling at itself. On the other side is the mere Earth, treated as dross. Nature, said C. S. Lewis is the material world after all of human worth has been wrung out of it.[4] From this prevalent viewpoint, other-than-human nature is without inherent value. There it is, just so much stuff for us to manipulate, exploit and transform, according to Francis Bacon, Adam Smith and Karl Marx.

Sickness in the Ecosphere is symptomatic of these wrong-headed beliefs. Battered Nature is insisting that a third possibility be explored. Beyond the individual and the society, though including them both, lies another substantial reality: the Ecosphere. It is no contractual expedient, not a prudential organization, but a *supra*-human reality that transcends in time and space the dialectic of individual/society, offering a higher-level synthesis and presenting new possibilities to the question of who in the world we are. It tells us that people are ecological beings—parts of the Ecosphere's sectoral ecosystems—and that people's rights and social responsibilities must be harmonized to that indisputable fact. Social, political and economic order may be only a matter of agreement between autonomous individuals but ecological order follows laws of a higher level—not those enacted by one species for its convenience.

In educational theory this means elevating ecology to a position of leadership over the unruly social/individual duo that, straining in opposite directions, generates social disharmony and paralysis. The goal of human ecology—the search for a healthy people-planet relationship—must set the direction and be the arbiter of conflicting "truths" that knowledge of the collectivity and of the individual set against each other. Human ecology has much to contribute on the subjects of progress, personal freedom, aesthetics, private property, social welfare, population growth, national aspirations and pursuit of the arts and sciences. The basic goal of a liberating education—understanding what it means to be human in a living world—needs first attention in the curriculum. Enlightened action will follow.

What is to be done?

Commentators on the state of the world inevitably are confronted by the question: "What must we *do*?" The response of Ivan Illich is not entirely a cop-out: "I've told you the problem and now you want answers too!?" Prime time should be spent getting our thoughts straight before rushing out to save our species and the world. Yet we are prone to act before we think, convinced that doing something, anything, will eventually accomplish what needs to be done. If the first attempt misses the mark, at least it will reveal the required curative action the second time around. As an example, many believe that the meaning of the slippery phrase "sustainable development"—that suggests we can have our cake and eat it too—will be revealed by trying this and that rather than by hard thought. Since an agreed-on definition is lacking, we will discover in practice what it means.

The naive belief that action leads to straight thinking is contradicted by the fact that theories and hypotheses guide behaviour. Actions unexamined in any profound way simply follow the network of conventional thoughts and paradigms we call culture. By this route, the solutions discovered for novel problems turn out to be the same old tired ones. Which is why science cannot be trusted to prescribe cures for environmental sickness.

Science and its methods are so pervasive in our culture that they have come to be the exemplars of how we ought to think and act, "objectively and rationally." Other equally important modes of knowing—artistic and affective—have been blurred by science's prestige. It has offered young people the promise of selfless service and a high calling in the search for Truth. The status of science, along with its language of mathematics, is pre-eminent in educational curricula. In this pre-eminence there lies a terrible danger.

Science is a cultural pursuit, meaning that it is rooted in our western philosophical tradition. It both reflects and reinforces society's dominant values and goals. As a body of confirmed knowledge and as a way of getting knowledge, science has been human-centred not world-centred, homocentric not ecocentric. Its traditional agenda, hidden because unquestioned, is service to humanity no matter what the costs to the rest of creation. Its traditional goal is manipulation and control. Its methodology is intensely masculine, aggressive, competitive and non-affective. The answer to the question, "Why are fewer women than men attracted to science?" is staring us in the face.

Science is instrumental, a means harnessed to human goals. As long as the goals are wrong, science will be part of the problem not the answer. Because of its inertia, because of its vast literature of control/manipulation, because of its focus on remaking nature for human convenience rather than on understanding and working cooperatively with nature for sustainability, science itself needs to be rescued and renovated. Science's new direction must be the caring search for knowledge that is harmonious with a larger-than-species vision, ecocentric not homocentric.

Ecocentric Education

Ever since the 1960s, educationists have recognized the ballooning importance of a vague subject of concern: "the environment." How to deal with it in an established curriculum? The easiest and least disruptive way to meet the challenge has been to encourage all teachers to take an interest in "environment," inserting "environmental concerns" into their classes. Not a bad idea, except that the all-important subject—human ecology—still lacks focus. It is as if the importance of calculating were to be recognized in all subjects without any special classes in mathematics, or as if the importance of budgeting were to be recognized in all government departments without establishing a leading department of finance. The life-preserving importance of the ecocentric idea requires that it erupt into the school curriculum as the central subject to which all others are harmonized.

A core curriculum reflects the consensus that certain subjects have paramount importance. The core assures that everyone is exposed to the great truths of a culture and to the means of communicating them. My contention is that in today's world *human ecology* must be the single essential core, approached through attention to natural history before cultural history, ecospherology before geography.

This is not an argument for ecology misconceived as a sub-discipline of biology, studying populations and communities of plants and animals in laboratory and field. Such traditional "ecological science" is useful insofar as it introduces young people to the natural world and its marvels, trivial insofar as it perpetuates the myth "life = organisms" and the myth that truth is only the apparently objective and quantitative. Human ecology is far broader and deeper, considering all people and their cultures as components within an evolving ecological system, the Ecosphere. People as parts must serve the whole—or bring on their own pathological demise. Humanity's arts and sciences must harmonize with the greater surrounding reality—or speed the race to an ignoble end.

Conclusion

"Education For A New World View" suggests a fresh purpose for education in the Age of Ecology: to lead learners away from the ignorance of their species-centred universe to a wider more invigorating view of planet Earth as a creative being, inviting humanity's cooperation and care. Such a purpose, necessary to stem the tide of environmental deterioration, challenges education at its deepest levels. It cannot be met simply by increasing the ecology content of the biological sciences nor by ingeniously exploiting those occasional sympathetic references to Nature within the arts curriculum. The goal—new to us and strange—requires a radical reversal of emphasis, shifting the spotlight away from humanity and focusing it on the Earth in which humanity exists as part.

H. G. Wells' opinion that education is always in a race with chaos seems amply verified today. We desperately need education as a new way of knowing—a new kind of science, one equally attractive to women and to men—whose goal is compassionate service to the Ecosphere and to humanity as a symbiotic part within it. With such science providing the "new world view" and lighting the path, the humanities and social sciences might also be reoriented ecologically. On the other hand, why should they not lead the way, giving the cue to the natural sciences?

References

[1] Orr, David W., "Is Conservation Education an Oxymoron?" *Conservation Education* Vol. 4 No. 2 (1990) 119-121.

[2] Neatby, Hilda, *So Little for the Mind* (Toronto: Clarke, Irwin & Company Limited, 1953) 232-233.

[3] Mungall, Constance, and Digby McLaren, eds. *Planet Under Stress* (Toronto: Oxford University Press, 1990).

[4] Lewis, C. S., *The Abolition of Man*. ed. Geoferey Bles, (London: The Centennial Press 1946).

MOW, MOW, MOW THAT GRASSY LAWN

I was mowing my lawn one fine summer day, like everyone else, when a young gal came along and waved at me. I could see she was a smart one, probably a this-year's grad, so I throttled down my noise-maker to have a little chat. She cast a sharp eye over my close-cut carpet, then asked: "Know what's America's biggest crop?"

"Yeah, sure," I said, pounding my ears. "It's wheat, or corn, or soybeans."

"Wrong," says she. "You're cutting it!"

Then she tells me the total area under turfgrass in the US of A. is a whopping 40,000 square miles, most of it in home lawns. Canadians try to keep up. Although many grow flowers and vegetables, which takes some intelligence, the majority spend their time whacking off grass tops.

"It works out," she said, "to about thirty hours of zombie mowing a year for every man, woman, and child in the nation."

I'd been reading the *National Post* so I knew the right answer.

"Yes, but lawns are good business," I said. "The turfgrass industry has annual sales of $25 billion, and that ain't hay!"

Little joke there and she snorted.

"So money and jobs make it OK? Next you'll be telling me it's good for the environment too?"

She's not the only one that's taken biology at school, so I let her have it.

"This grass," I said, giving it a kick, "is turning out oxygen even as we speak. A nice, neat 50 x 50-foot lawn makes enough oxygen for your ordinary family of four!"

Did that knock her back? No, she snows me with figures about how much oxygen the mower engine, the fertilizers, and the lawn pesticides use. Turns out that, on balance, these grassy yards are sucking up oxygen. If nuclear families depended on their lawns for breathing, they'd soon be nuked.

But I had an ace up my sleeve. "It's part of our culture," I said proudly. "It's a Canadian institution. You've got to admit that there's something soothing in the rhythmic pounding of the old 2-or 4-strokers in everybody's yard. When one of 'em starts up it reminds everyone else to start up. Just like frogs in the spring. It's a matey national chorus, plus a touch of the perfume of the Alberta oilfields."

I could tell by the gleam in her eye I should have yanked on the starter cord.

She pulls out a book that says an hour of pushing or riding a power-mower releases as much pollution as driving your car three hundred and fifty miles, which is why California is chasing those mowers off the lawns; that the average acre of yard grass gets four times as much poisoning as the average acre of farmland—which is still enough to make the farmers sick.

Who needs more bad news? I just tuned her out and went back to cutting the grass mindlessly. Well, not exactly *mindlessly* because I was thinking: "If I let the grass grow, it would smother all the dandelions, and then what would I do with my can of poison-spray?"

So I better keep cutting her down.

4. HOMO ECOLOGICUS SHORT ESSAYS

THE TRUE VISION

Some years ago, in my forestry days, I was temporarily part of a government window-dressing ploy called the federal "Forest Advisory Council." While we partied from place to place across the country, our main duties, the Minister occasionally reminded us, were to provide clear-headed and far-sighted advice on forestry policy, balancing the presumably muddle-headed advice of his Deputy Minister and subsidiary crew of bureaucrats.

This formidable task was not rendered easy by the fact that, apart from two academics and one non-governmental organization representative, the Council was composed of industry big-wigs and conservative buddies of the Minister. To the challenge of changing a deteriorating status quo, they brought only a vague, oft-repeated mantra: "Forestry needs a new vision."

The phrase "a new vision" is the artist's leading light, the goal of the creative, the hope of the reformer. Even the scientist has been known to entertain the notion that the direction of human society is not completely set by physical causation, by the push of the past, but can be influenced by the vision of an imagined future: a pull from in front. This concession to free will, to the validity of purposively pursuing novel social goals, is made reluctantly (if at all) by the "hard" sciences. It is more at home with the "softies"—the environmental and social sciences, the arts, the humanities.

As ecological and social crises blossom world-wide the cry goes up for a New Shared Vision, a new overarching theory or paradigm that will bring all peoples together, setting before them understandable and inspiring goals.

Now and then I look at an e-mail list called "Ecopolitics" where various schools of environmental theory—green environmentalists, social ecologists, ecofeminists, deep ecologists—struggle to find common ground. Mostly they agree that getting the underlying theory right matters, that people can only work together constructively when they share a common faith and philosophy. On this point the deep ecologists argue, wrongly I believe, that the necessary social reforms would be instituted, regardless of fundamental faith differences, were people to accept and act on one main principle: all organisms are inherently valuable. This exception noted, agreement is general that environmental problems are the outcome of the current faulty world view. It is not just forestry that needs a new vision. "Humanity needs a new vision!"

A famous name in the ranks of social ecologists is Robert Constanza who a few years ago teamed up with other prestigious conservationists to write an article titled "Managing Our Environmental Portfolio" (published in the journal, *BioScience*). The new vision he espoused is that of the prudent biblical steward who invests his talents profitably in Earth to ensure a good return and, presumably, to secure a comfortable retirement. "Be ye good stewards" seems to be the current popular thinking on the greening of Earth. To it the theologians, and humanitarians from Mumford to Bloom, add their vision of people improving in attitudes toward one another, becoming more tolerant, kinder, more creative and innovative, more artistic. Excellent goals but hardly a new vision. Such urgings segue too often into an obsession with spiritual growth—an ancient trajectory that misses the outward point.

Visions may seem to come from the "within," from dreams and inward ponderings, but tracked to their source in images and words they are securely anchored in the "without."

The very word "vision" is based on the seer, on seeing, on comprehension of the needs of humans—not just as individual and social beings but, far broader, as beings of the Earth.

Look! The new vision surrounds us in the trees and the flowers, in the clouds and the rivers, in the mountains and the sea. Its name is symbiosis, which means making a life together. The new vision is *out there* and always has been. It is the spring of inspiration, the source of whatever good has been discovered *within* the human mind.

"HOMO ECOLOGICUS"—A WISER NAME

A magazine ran a little contest asking what one might say diplomatically to an actor friend who had just turned in a ragged performance. A winning answer was the emphatic, "Good isn't the word for the way you acted tonight!"

Lately, reading the daily papers, I've been thinking, "Good isn't the word for the way we're acting today!" Human carryings-on seem completely divorced from any sense of where we've come from, who we are, and what keeps us going. City lights and indoor living are partly to blame.

Everyone should spend a few nights every year in the countryside under the stars, vivid reminders of the immense universe wherein our little lives are lived. Those twinkling celestial lights—once close-packed together—have been flying apart for billions of years. Now and then an exceptionally bright one appears in the sky, an exploding "supernova," and in its fiery furnace the heaviest atoms are formed. This explains why Earth's magnetic core is composed of heavy metals, many of them intensely radioactive, the partial fallout of a star that went supernova four or five billion years ago. The event marked the birth of our solar system with its one special life-giving planet.

Earth's density decreases from the core outward. The continents are composed of light-weight rocks floating on the heavier ones beneath. Lighter still is water that fills the ocean and continental basins. The gases, lightest of all, form an outer shell: the atmosphere. From the buoyant stardust at Earth's surface, organisms magically arose—twenty or thirty million different kinds in today's world—among them a featherless biped with a big brain, self-named *Homo sapiens*, "Man the wise."

The Wise One knows his body chemistry is derived from air, water, and soil. The average body floats in water, showing the absence of deep-Earth components. Like its relatives in the plant and animal world, the human body/mind is a product of cosmic froth—the light-weight substances that lie about on Earth's surface. People are made from clean air, clean water, and clean humified soil—all those things that exist in Earth's green surface film. To these their healthy bodies are adapted.

Homo sapiens ignores this fundamental knowledge, secure in the belief that human errors can be absolved by faith, or cured by science and technology. Thus the Wise One makes a virtue of mining and releasing into air, water and soil the many poisonous materials that used to be safely sequestered underground: heavy metals, radioactive minerals, long-chain hydrocarbons such as coal and oil. Nor does it deter him ("her" hasn't got into the act yet, though she is learning his ways fast) from creating a multitude of chemical poisons such as the organochlorides—DDT and PCBs among them—and loosing them into the surface life-space.

The Wise One believes that nothing but good can come from having as many children as one's income can support, ignoring the fact that each cute baby grows into a consuming adult who demands food, water, shelter, and a multitude of energy-consuming, pollution-producing machines in order "to live the good life." According to the geological record no large animal has ever existed in numbers approaching the six and a half billion humans enthusiastically reproducing toward the goal of eight or nine billion a generation hence. As booming populations crowd other creatures to extinction, stealing their habitat niches, cosmetic campaigns are launched to save endangered species, to save the grasslands, to save the tropical rain forests, to save the coral reefs, to save the soils and the seas.

An article of faith rarely questioned is that all things on Earth (and indeed in the universe) have been placed there expressly for the use and pleasure of *Homo sapiens*. Thus all fertile soils are to be planted to crops, all forests are to be logged and converted to plantations, and all streams and lakes must prove their worth by yielding fish, hydroelectric power, or recreation. Mammalian relatives on land and in the oceans are there to be killed for food, for sport, or for science. The worth of everything on Earth, except people, is tested by the question, "What use is it to us?" If the answer is "no use," dismiss and forget it.

Clearly *sapiens* is not an accurate descriptor for humans. A new name is needed, and a candidate that springs to mind is *Homo ignoramus*. Although factually correct, it might prove too crushing for the delicate egos of scholars and scientists. *Homo islandicus*? *Homo disconnecticus*? *Homo navel-gazibus*? All perhaps too negative.

Ask then, what name would be true to humanity's Earth-bound nature, a beacon to counter and correct the narcissism of the species? What might shine a light on the intellectual/emotional blind spot of the intelligentsia, those who set and confirm the cultural norms that guide politics, economics, science, and the arts today? Otherwise put, what is lacking in the contemporary thinking of the great humanists, the influential homocentrists?

Their chief deficiency (to my mind) is affection for the living Earth. They lack a clear understanding of humanity's absolute dependence on the green world, an understanding that goes far beyond a "concern for environment" or a "willingness to conserve biodiversity," to which many now subscribe.

Needed is a supra-humanist sense of people sharing the planet with other life-forms rather than over-powering them, letting evolution have its way for the next million years in the interest of evolving higher levels of sensitivity and intelligence. This vital ecological intuition the Mumfords and the Bronowskis, the Fryes and the Blooms, the Postmans and the Wilsons, uniformly lack. Perhaps, significantly, the Atwoods, Spretnaks, and Butalas come closer to the mark.

Homo ecologicus correctly describes humanity's unrealized status. The name suggests the truth that we are animals intimately related to all that surrounds us. We are Earthlings, parts created and sustained by the planetary matrix. *H. ecologicus* could be the inspirational inscription on signposts for the future, pointing toward the path of genuine progress, the path not yet taken.

If "good" is not the word for our performance to date, can a new name encourage a better act? It is well worth a try.

BRAINS ARE THE PROBLEM?

Sitting at my open back door facing south, I'm soaking up the February sunshine, blinking my eyes at the blue skies and snowy evergreens, wondering, "How come everyone knows we're tormenting this beautiful world, yet we can't seem to stop?"

Trucks and cars buzz by, warming the Earth as they trail the perfume of partly burned hydrocarbons. Eastward on the forested mountainside the first ominous, rectangular clear-cut patches are visible. Then I think that soon, this year or next, the provincial Health Officer will be ordering the village to chlorinate the local water supply. Pure and sweet at the moment, our spring-fed water is officially described as "raw and untreated."

Anyone who listens to the international news, or reads an international newspaper, knows that horrific "environmental problems" are coming to a head everywhere in the world. Human wear-and-tear on the planet is global. Something's radically wrong and we can't seem to fix it. Why aren't the world leaders, the Prime Minister, the premiers, and city, town and village councils doing something about it?

In recent years women have been trying to educate me, pointing out that all our leaders are guys—a group to which, anatomically, I have reason to believe I belong. They tell me everyone's big brain is divided into two lobes, right and left, that give different but important messages. In general women have a healthy balance between the two lobes, while men tend to be strongly left-brained. Because men run the political and economic show, maybe the root problem is that only half a brain is calling the shots?

The right lobe is the intuitive side, the artistic side, the gut-feeling side, giving emotional reactions to every experience. "Trust your intuition!" means "Listen to the right lobe!" The right lobe is always signaling likes or dislikes, what's good and what's bad, comfortable or uncomfortable, beautiful or ugly. It says to me, for example, "I feel there's something wrong with clear-cutting a wild forest and putting a tame plantation in its place," to which the left lobe replies, "Feelings, shmeelings! Just give me the facts!"

For the left lobe is the rational side whose job is to provide reasons for conscious experiences. It says, "Look, you can see that trees, and rain, and humans are separate, unconnected things. That's fact! So clearcut the forests, carefully of course, and it won't much affect the landscape, or the water, or the wild plants and animals, or those most important separate things, people."

In expressing reasons, the "look" and "see" language is often used. Sight is our strongest sense, and sight works by separating the world into individual things. Put another way, sight breaks the entire world of interconnections into separate pieces, and the left brain says, "Of course, that's reality!"

Now it turns out that a lot of men, especially the smart ones, are away out in left field, short-changed on the right-brain side. There's even a name for the affliction in clever men like Albert Einstein. It's a form of autism called the "Asperger Syndrome" (ASS) and men are four times as likely to have it as women. Here are Bill Gates' symptoms: "Poor social skills, lack of eye contact, a monotonous voice, a prodigious memory, and a tendency to rock back and forth during business meetings."

Maybe we need to trust our gut feelings more? Maybe there are too many ASS's in charge? Maybe we need more women running the country? I've been rocking back and forth, wondering which would be best?

My, how warm is the early spring sun, lighting the cerulean sky, sparkling on the snowy evergreens. Somehow seeing and feeling come together on a day like this in the one word: good.

FROM THE GROUND UP

Behold your average or below-average gardener, with moderate success growing peas, beans, broccoli, carrots, onions, kale, and especially never-fail lettuce, in several raised beds between a row of raspberries on one side and strawberries on the other, ineffectually striving to protect the plum tree from aphids and leaf-rollers, the maturing apples from canker worms. But though I have much to learn about growing food for myself and others, I think I know why it is an excellent idea. Gardening puts our body/minds in touch with this miraculous planet Earth from which our body/minds are made. It is also preparation for the coming day when regional self-sufficiency—ecoregionalism—especially in food will be a top priority.

About one hundred and fifty years ago, Darwin pointed out the obvious fact that we humans are one of Earth's mammals, a primate closely related to the chimpanzees and other great apes of Africa and Asia. Despite a mountain of evidence since then that we share the same genetic materials, many people resist thinking of themselves as one of Earth's creations, closely related to the rest of the animal kingdom and, at greater distance, to the plant kingdom. Haemoglobin and chlorophyll, blood and green sap, are next-door chemical cousins, easing the way for many of us to be politically Green. What a marvel that our beginnings, way beyond animals and plants, can be traced to the kingdoms of microorganisms from whose confederacies our physical bodies are made. At base all organisms are composed of Earth's air, water, and soil. Humans from humus, as ancient languages tell us.

Instead of accepting this reality, many have emphasized how different we are from the rest of creation—you know, special, exceptional, a little lower than the angels, God-like and made in His image, therefore outside the rules of Nature. This falsehood underlies widespread mistreatment of the Earth, causing environmental problems that are really people problems. A healthy view would perceive these troubles as resulting from a lack of harmony between human activities and Earth's processes, a warning reminder that as parts of Planet Earth, as totally dependent Earthlings, we must learn how to live amicably with Mother Matrix or she will disown and reject us.

The history of human evolution tells important things about ourselves. Distant ancestors came from Africa where for several million years they evolved like Tarzan and Jane, happily swinging in the trees as little girls and boys still like to do. As forests shrank they evolved into upright walkers/joggers suited to the flowery savanna with grassy expanses and scattered trees—the way we like to arrange our city parks for the convenience of walkers and joggers, and to green-carpet out homes with floral wallpaper. Then sometime in the last million years or so these clever ancestors (who by one hundred thousand years ago looked just like us) migrated out of Africa, and with fire and stone-tools invaded first the warmer parts of the continents—Asia and Australia—then Europe, and in the last twenty thousand years the Americas.

Agriculture, the relatively recent method of subsistence by turning the soil upside down then planting food crops, was only invented about ten thousand years ago, after the last great glaciation. That invention changed most of the wandering foragers into settlers, who stayed in place to guard and harvest their food plants. Agriculture fostered the growth of towns, then cities from which the word "civilization" comes—as if rural life cannot be civil! Until the Industrial Revolution, that arrived only one hundred and fifty years ago in North America, most everyone was still making their own clothes and growing their own food.

For 99.999 per cent of human evolution the natural healthy way to live physically and psychologically has been in the outback, either foraging for food like the aborigines of western North American, or practicing gardening and small-scale agriculture like their eastern cousins. Through long evolution, while living on and from Earth's land and water, the human animal has been tuned genetically to this special living planet with its moon that pulls the ocean tides. The way our body/minds, organs, and cells function cannot be separated from the way the Earth-Moon duo behaves as together they loop around the Sun every three hundred and sixty-five and one quarter days. All our body/mind rhythms of sleeping and waking, of eating and fasting, along with rhythms of hormone releases, reproductive functions, the aging process, etc., are matched ecologically to the outer cycles of day/night lengths, cycles of the seasons, cycles of the years.

Like our relatives the bears and raccoons, we northerners evolved in a seasonal pattern, fattening up in the summer and autumn, eating grains, vegetables, and fruits, and then sleeping it off over winter. Skinny from hibernation but rarin' to go, our Mediterranean ancestors emerged from their caves around Valentine's Day to celebrate spring and return of love goddess, *Juno Februata*. In the more northerly climate of England the spring festivities were marked later with a fertility dance in early May around the May-pole.

How differently we spend the dark nights of December, January and February now, all warm and lighted up in our dry-wall caves! Thanks to electricity we have created eternal summer indoors, fooling our bodies further by importing summer foods from the tropics and sub-tropics. Each year we fatten up for a cold, dark winter that never comes. Thank Edison for making today's selling of diets so profitable. Now we can dance around the May-pole at any time of the year.

Electricity is but one example of the way we have given our Earth-smart bodies a heap of wrong signals that adversely affect health. No other animal species, as far as we know, sets out to defeat the cycles of the

Moon, Earth, and Sun to which all living bodies have been adapted. Even the dumb plants wake themselves by somehow counting temperature sums in the spring, and go to sleep by counting the decreasing daylight hours in the fall. If Martians were watching us, they would have our Western culture tagged as insanely unecological. They'd be saying, "We call ourselves Martians because we're made from and fitted to Mars. Don't these humans realize they are Earthlings, from Earth? How little attention they pay to the signals of their source and support!" That cannot truthfully be said of dedicated gardeners who must attend to the changing seasons.

Gardening, growing food and flowers, satisfies the two human desires to do pleasurable delightful things and to avoid dangers, discomforts and pain. Gardening gives pleasure because it fits our nature and, contributing to self-reliance, it is a defence against future shock.

(1) Gardening is pleasurable

Composting, working the soil, growing food plants and flowers, puts us in touch with reality, with Earth and its cycles that are our source and support. Gardening calls into use all senses and all muscles. Not only growing food but also storing it and preparing it ourselves for eating, is a step in the right direction. In growing and eating our own plant food, raised in our own compost and soil, we are re-establishing ties with Earth and satisfying a natural desire. At the same time, as social animals, the instinctive desire to care for our own well-being and that of our family and friends is satisfied.

(2) Individual and Community Gardening Hedges Against Future Shock

While we are learning to garden better and co-operatively, we are hedging against failure of the prevalent economic system. The hydrocarbons underground—gas and oil—are limited in quantity, and they are non-renewable. Scarcity in the future will drive their prices sky-high, and some experts expect a crisis within the next decade or two. Forget the glib talk about future hydrogen economies, or nuclear economies, there is no cheap alternative to oil and gasoline, so the price of travel and the price of imported food will soar. Trucking food to small communities will be more and more expensive. Local food security is good insurance against an uncertain future.

Local self-reliance is also a small contribution to preventing pollution of air, water, soil, and plant food. Humanity is in danger of smogging itself and its food supply into a permanent state of illness, along with irreversibly changing Earth's climate. The internal combustion engine that powers most vehicles is the main culprit, and ways are being sought to limit its use. The average commercial food item in North America has travelled thirteen hundred miles between its place of production and the dinner plate where it is consumed. Reducing transportation of life's necessities will help to avoid harm to the Earth home.

Ecoregionalism, striving for self-sufficient communities within large natural regions, is probably the only goal that can lead to long-term sustainability in land and water use. It is certainly the only goal capable of planned recycling, for global trade in food stuffs can never put back into agricultural soils what has been taken out and shipped far away. Further, it is wise to rely on technologies providing food, clothing, shelter, energy, that are operated as much as possible by people in local communities, people we get to know, people we can work with. It is unwise to rely for essentials on far distant supplies that are beyond control and always at risk from market swings, climatic disasters, and today's terrorism. If that sounds paranoid, remember, "Sometimes only the paranoid survive!"

THE EAGLE'S-EYE VIEW

Although I am mechanically inept, or in less judgmental language, "manifold-challenged under the hood," I have had a lifelong interest in small flying machines, especially helicopters and Cesna planes. They intrigue me not only by their gravity-defying powers but, more, by their usefulness as rapid transport from place to place over North America's expanses of grassland, forest, peatland, and tundra, always revealing spectacular patterned landscapes. How fortunate I have been to look down on so many varied scenes, their marvels largely unappreciated by the general public because seldom viewed, bird-like, from the air.

A fortunate naturalist, I have been sky-high over most of the north, gazing in wonder at the lineaments of Earth whose fundamental characteristics are expressions of the underlying bedrock. Slung around Hudson Bay like a broad horseshoe, ancient pink-and-grey crystalline rocks of the Precambrian Shield are exposed. This is the core and strong fabric of the North American crustal plate as it sails majestically westward into the Pacific Ocean at several centimetres per year. Sedimentary rock formations overlie the margins of the Shield, concealing its presence under the rest of the continent. They are frequently banded in narrow layers like tree rings, each the yearly accumulation of clay, silt or sand at the bottom of shallow seas which, from time to time, invaded the continent. Often they entombed sea creatures whose fossil remains give character to the building stones of our cities.

In Earth's recent geological history, great glaciers formed and slowly flowed over the exposed Shield and its neighbouring sedimentary formations, eroding and mixing the surface materials to form glacial till, deposited as moraine on the underside of the ice and at its melting margins. Such landforms comprise today's settled landscapes: gently undulating till plains; hilly belts of hummocky moraine known as "pothole country"; sinuous eskers like railway embankments marking the gravel beds of former rivers on, in or under the ice; the curving strandlines of ancient ice-margined lakes, long since drained, whose silt and clay bottoms constitute today's best stone-free agricultural lands.

On these foundational features lie innumerable multi-coloured lakes and the lively patterns of plant communities whose summer tones are the tans of grassland, the reddish brown of peatlands, the soft greens of deciduous forests, and the darker greens of coniferous forests. Everywhere, in wooded terrain, one can see from the air a mosaic pattern of even-aged forest patches that differ slightly in green shading according to species, composition and age, each the mark of a lightning-ignited fire—the renewing agent of most northern vegetation types.

Seasonal changes in the vegetation make striking patterns when viewed from aloft. One marvellous experience I vividly recall was the flight over a vast mature aspen forest in the wilds of northwestern Saskatchewan. It was a blue-sky, late-September day, the air crisp and clear, and the leaves of the trembling aspen had completed their seasonal change from green to a dazzling yellow. We flew several hundred feet above the unbroken forest canopy, just high enough to make it seem a rich continuous surface. Dazzled by the brilliant reflection, I gazed in a trance at the luminous carpet as it approached and passed beneath us, the quivering leaves atop the nubbly tree crowns adding a subtle animation to the texture below. Westward we flew over leagues of glowing forest and then, off on our northern side an unexpected sight: a scarlet patch of forest shining in the sun, a ruby island in the yellow sea. As our plane-shadow moved past the mutant clone—a crimson maternal tree surrounded by her similarly tinted daughters—the grand canvas assumed a Van Goghish cast of primary colours. The shimmering leaf cover with its yellow-red contrast, set against the Mediterranean-azure sky above the horizon's rim, painted for a moment the "Riviera of the North."

And "Riviera of the North" is the description often applied to the Shield country west of Hudson Bay that coincides with northern Saskatchewan, Manitoba, and much of the Northwest Territories. For a brief spell the immense tracts of wooded northland, home to the Cree and the Chipewyan nations, enjoy a sunny Mediterranean climate. Here in the land of lakes, peatlands, rock outcrops and forested sandy soils the summer days are long and warm, rainy spells are few, and the forests on thin-soiled uplands quickly dry to the "high fire-hazard" rating. Then the air is sweet with the smell of parched pine and spruce needles, the yellow and grey-green lichen mats on the woodland floor crunch underfoot, and every stroke of lightning from late-afternoon "dry thunder storms" ignites a tree and the leaf mat at its base. A patch is burned, and on it a renewed birch/pine-/spruce forest soon appears.

Once I saw an extraordinary example during an October flight over Shield country on the way to Yellowknife in the Northwest Territories. We cruised along at an altitude of a thousand feet or so, mile after mile passing over the spruce/pine tapestry dotted with paper birch. This was woodland rather than closed-crown forest—a splendid park-like setting where each tree was spaced apart from its neighbours, no crowns touching. The preceding weeks had been windless and the autumn leaves of the birch trees had fallen without drifting, creating within the darker matrix a design of bright circles each centred on the white upright line of a birch trunk. The effect of the repeated pattern—golden plates set in a green tablecloth—was magical. Here was landscape art on a grand scale, composed and coloured with a palette of greens and golds, streaked with the light and dark verticals of the deciduous and evergreen tree stems.

Would that everyone could be so uplifted, raised sky-high to admire the beauty of natural landscapes, to see their pristine splendour that cleanses the too-much-inturned vision! Thus might be justified, at least in small part, the infernal internal-combustion engine.

BIRDS

Sixty million years ago the dinosaur herds shrieked and snorted, stampeding through marsh and river valley as the sky grew dark and bitter. An asteroid had smashed into the Gulf of Mexico cloaking the Earth with sun-screen dust, releasing vast clouds of sulphur that fell as acid rain, destroying the lush greenery at the base of the food chain. And so the dinosaurs vanished?

Well, not quite. The small ones and their relatives survived. The

Western Skink and the Northern Alligator Lizard sun themselves on the rocky shores of Slocan Lake. In the southern corner where Alberta meets Saskatchewan the Short-horned Lizard, commonly known as the horned toad, scrambles through the dry sagey grass, while eastward in Manitoba the Northern Prairie Skink disports herself. We have half a dozen snakes, diminutive lizards who found legs more a bother than useful. And did you, perhaps this morning, breakfast on an incipient little dinosaur—the egg of a fowl?

Take a close look at yonder hen, at her cold reptilian eye, her scaly legs and feet. There, clucking disarmingly, stands a winged dinosaur. Be glad you are not the size of a worm, for though toothless today the "terrible lizards" live on as our feathered friends, tracing their heritage back to the Jurassic *Archaeopteryx*, a creature with a lizard-like tail, well-developed wings, and a reasonable set of dentures. From this came the ostrich, standing eight feet tall, the brontosaurus of the bird family and, at the other extreme, the moth-sized humming-bird which, like lizards, readily passes into a saving state of torpor when the nights are cold.

Just as our mammalian stock branched off from the "saurian" line millions of years ago, so did the birds. Mutual genetic material and an ornithological brain-stem—our *medulla oblongata* is the original "bird brain"—remind us that we are related in evolution and therefore, to some extent, united physically and mentally. Perhaps that partly explains our fascination with these strange winged creatures, our enchantment with the marvelous architecture of hollow bones and interlocking feathers that moves them through the air, their beautiful plumage patterns, their exotic courting and mating habits, and their attractive songs: Earth's first harmonic music.

As we kill birds by the millions every year through the agency of our totalitarian agriculture, by destroying natural breeding habitat in the north and over-wintering habitat in the south, by biocides to exterminate whatever seems a nuisance or useless to us, by night-lighted buildings, by overhead wires and pet cats, the fascination with birds and bird-watching steadily grows.

Setting out the sunflower seeds, rescuing little birds fluttering against windows, de-oiling the bigger ones trapped in spills of bunker-C, rehabilitating gun-shot hawks and eagles—these evidences of goodwill show a deepening sensitivity which, if nurtured, may yet turn humanity toward a more compassionate way of being in the world.

The exemplary story is told of a benevolent woman whose heart was torn every time she witnessed some heedless crime against the birds she loved. Although she cared for her mate, his rank in her affections gradually sank to second, eclipsed by her passion for helping suffering birds. Resentment grew in his bosom.

One evening, returning late from work tired and hungry, he found the kitchen table bare. His wife had rescued from the cold a half-frozen wren, which she had brought indoors and wrapped in a warm towel. Instead of preparing supper she was soothing the little wren before the fire.

The husband's temper snapped. This was the last straw, he had had enough! He ranted, he raved, he swore, until in mid-blasphemy she stopped him with upraised hand.

"Please dear," she said pleadingly, "not in front of the chilled wren!"

GENESIS

In the beginning was "the world," a cloud of stardust without form and void, and this was proto-Earth. And the dust was gathered into a spinning globe, a planet circling the burning sun, lighted first on one side and then the other, day succeeding night and night succeeding day endlessly, and it was good.

From a patch of Earth the Moon was formed, a satellite sphere circling the planet to which it was so intimately matched that it showed but one face. The moon lighted the night as the sun did the day, and it was good.

That which was heavy sank toward the centre of Earth while that which was light floated toward its surface. Most airy were the gases, clothing Earth with a transparent mantle, forming the blue sky-dome of day. At night the darkened firmament sparkled with a million stars and galaxies, and it was good.

Next in lightness the waters steamed forth from Earth and gathered in streams and oceans under the heaven. And the lightest rocks floated on the heavier rocks below, rising above the waters to form dry land. The Moon tugged at Earth, pulling the tides of the seas, and all of it was good.

Under Sun and Moon, the circling Earth brought forth a profusion of creatures made from its air, water and surface sediments: first the small and single-celled and then, by joining together, the many-celled organic forms. So from Earth came grass, and herbs yielding seed after their kind, and trees yielding fruit, and great whales, and the beasts of the Earth, and cattle, and every living creature that moveth, each after its own kind, and all of them were good.

Then Earth brought forth mammals and primates, and from them people. And the people gazed around them with wonder and said, "Earth is beautiful in its greenery, with plentiful food from the herbs yielding seeds, the trees yielding nuts and fruit, fish from the water and fowl from the air. Here are flowery meadows, sweet air, and clear running water. This Earth is a garden and here we live in paradise."

Then they bethought themselves why this should be, and they turned with questions to their Wise Men well stricken in age and fearful of death.

"Know this," said the Ancients, "that people are a special creation, nay, the summit of creation. All that has gone before in the history of Earth but prefigured God's great purpose: to make humans and give them immortal souls, making them godlike. God created man in his own image, in the image of God created he him; male and female created he them."

And on hearing this pleasing fable, everyone agreed that it must be "just so."

"Know this further," said the Sages, "that having created people in His own image, God would not place them in an unkempt world. Earth He made beautiful for the delight of His children. All creation is a paradise to be enjoyed."

And none on hearing, "For me the Earth was made," questioned its truthfulness.

Then the Wise Men sun-struck in the desert warmed to their prophetic task, telling how God had blessed the human male and female that He had created, saying: "Be fruitful and multiply, and replenish the earth, and subdue it; and have dominion over the fish of the sea and over the fowl of the air, and over every living thing that moveth upon the earth. Behold, I have given you every herb-bearing seed, which is upon the face of all the earth, and every tree yielding seed; to you it shall be for meat."

This fanciful story, cast in poetic phrases, was music to the ears of its hearers, and they spoke approvingly to one another saying, "Beauty is truth, and therefore that which falleth trippingly from the tongue and soundeth good must be true."

Thus were the people misled by their sages. Not mindful that they were Earth's children, they thought the Earth their own and proceeded to eat it up. Supposing themselves a special creation, they multiplied exceedingly. Believing themselves godlike and heaven their immortal home, they treated Earth and its other creatures as objects for their use, neither with respect nor loving care.

And lo the paradise of garden-Earth began to fade. The beauty of the seas and skies was dimmed. Adieu to companion animals and plants. The people found themselves more and more alone in a shabby world.

And the shabby world that they had created made them, in its image, shabby people.

WHO, ME SUPERSTITIOUS?

One Saturday the emissaries of a well-known religious sect came knocking at my door and handed me a little magazine whose front cover featured a black cat, a pair of dice, the signs of the zodiac, and the question in capital letters:

"SUPERSTITIONS, WHY SO DANGEROUS?"

I looked from the cover to the serious faces of the ladies, then back to the cover again and couldn't suppress a chuckle, at which one remarked, "You must have just thought of a few superstitions!"

"Indeed I have," I said, looking her smilingly in the eye.

Of course she didn't get it. She could never believe that handing out religious tracts from house to house is promoting a particular superstition. And I am sure that if I knocked on her door and gave her this essay she would detect dangerous superstitions compared to her own true faith.

Faith and superstition are alike in being based on beliefs for which there is little shared evidence, hence no logical proof. We are all superstitious in one way or another, believing because others believe. Given solid evidence to which everyone is privy, faith and superstition are transformed into something else—matter of fact.

Usually we think of our own faiths as sound and reasonable. Each one of us looks skeptically at the faiths of other cultures and subcultures, usually writing them off as baseless convictions, as superstitions. An analogy is the question: who among the speakers of English have accents? Of course we in western North America speak the pure language; everyone else has an accent! Just so, within each culture and sub-culture, ours is the true faith.

The conversation with the missionaries started me thinking. If everyone is a carrier of some sort of faith, of some species of superstition, does that mean that one is as good as any other? Should we therefore be tolerant of all the strange beliefs amongst us? Or is there a way to order the spectrum of faiths, theories, superstitions, as good, bad, or indifferent?

Well, yes. One test is to ask what a faith's likely outcome would be if everyone embraced it and acted on it. Would it be good for planet Earth? Would it be good for the thirty million companion species that inhabit the globe? Would it be good for all people? Or would it do no more than alleviate the anxiety of its practitioners? Notice the order of importance in which the questions are asked. By this test practically every known faith/superstition comes up short.

Like all faiths, superstitions are especially dangerous when their original meaning of what "stands above" (*super/stitio*) the human race, what surpasses it in importance, is unrecognized and neglected. Clearly the twenty-first century needs a new faith/superstition with an outward focus on the life-giving and life-supporting Earth.

GENDER-NEUTRAL LANGUAGE

Anyone can avoid third-person-singular gendered language if they try hard enough, because if *anyone* aren't a plural noun it's high time they is.

The gendered orientation of language is important. Mary Daly opined that as long as God is male, male will be God. But sometimes limited choices are offered by language. Words fail us. For example, little choice of terms is available to describe humanity's love affair with itself. Shall we describe ourselves as "anthropocentric" or "homocentric?" These, like the parallel terms "anthropomorphic" and "homomorphic," are not in their usual senses gender-neutral.

Some have asked why I prefer to describe humanity's navel-gazing preoccupation with itself as "homocentric" rather than the more usual "anthropocentric"? Well, "homo" is slightly more gender-neutral than "anthropo," as we shall see.

The Greek "*anthropo*" means "man" and its relative "*andro*" means "male." The Latin "*homo*" also means "man." "Human" is from the Latin "*humanus*" defined as "akin to homo." Females, comprising 51 per cent of the race, are slighted by these manly terms.

Patriarchal English-speaking society has only one gender-neutral term for all of its members, viz. "people," from the Old French "*pople*," in turn from the Latin "*populus*" meaning "nation, crowd, or race." Unfortunately the terms "peoplecentric" and "peoplemorphic" are clumsy. They do not fall trippingly from the tongue, which condemns them as replacements for "homocentric" or "anthropomorphic."

But perhaps the word "people" can be abbreviated to the popular "pop" (as in pop music or pop art) and used as a gender-neutral combining term? Instead of the usual, "The gorilla is an anthropoid (man-like) ape"—an egregious insult to female gorillas—how about, "the gorilla is a popoid (people-like) ape"?

Such a neologism might pass, except for the long-established use of "pop" as an affectionate name for dad. On first encountering "popoid ape," many would undoubtedly jump to the conclusion that it means "an ape, like father." As a generalization this might, in some instances, be unjust.

The ancestry of words provides a solution. Although the Latin word "*homo*" means "man," it is derivative from the Sanskrit "*ghthem*" meaning "Earth" from which came the Greek "*chthon*" and in a roundabout way our word "humus." Dust thou art to dust returnest. So "homo" is literally grounded in Earth which used to be considered female (*Gaia* in Greek, *Gaea* in Latin) and is still, at times, fondly and appropriately called "Mother Earth."

The prefix "homo" therefore reminds us that we are dust, that human is humus is Earth is Gaia. In this sense the scientific name *Homo sapiens*—modestly adopted by humanity to describe the sagacity to which it aspires—lends itself to translation not as the congratulatory "Man the Wise" but, more down to Earth and heterosexually appropriate, as "Smart Humus" or "Clever Dirt."

The relationship between the words Homo, Humus, Earth and Gaia, puts the race in touch with its origin. Humans are born from Gaia, from the Earth life-source, rooted from birth to death in the mother/matrix, which perhaps moderates the sting of the bumper sticker: "Grow Your Own Dope; Plant a Man."

Further, by fortunate coincidence "*homo*" in the original Greek is the counter to "*hetero*" and means "same, equal, like," as in the words "homo/genize" and "homo/sexual." Thus *Homo sapiens*, translated with both a Latin and Greek twist, implies a species intelligent enough to recognize both the alikeness of its members, regardless of colour, creed, sex, or gender orientation, *and* its close ties to Earth—the creative source and support of all. When shall we see at universities the first Department of Homology?

So, my male friends, if one must make a choice between homocentric and anthropocentric, between homomorphic and anthropomorphic, between homogenic and anthropogenic, between homopoid and anthropoid, the first term of each pair is clearly preferable.

And when you use the favoured "homo" adjective in mixed company be assured that the eyes of perceptive women will warm and brighten as they recognize in their midst the outcome of ten thousand years of slow evolution: a SNAG—a Sensitive New Age Guy.

❧ REFLECTIONS ON MOTHER'S DAY

Mother's Day rolls around yearly, celebrated in North America on the second Sunday of May, and in the UK somewhat earlier—on the 4th Sunday of Lent, called "Mothering Day." Thus, one day annually, the English-speaking world acknowledges the presence of genetically XX Mothers, before turning attention back to the real XY world of the Non-Mothers—where truly important things get done.

It was not always thus. Diggings in the archaeological ruins of Crete, of ancient Sumeria in southern Iraq, and at Catal Juyuk in southern Turkey, have revealed signs of early mothering societies—where fathers were the ones who got the one-day-a-year treatment. Little goddess figures, with big breasts and sturdy hips, show not only who was recognizably in charge of fertility and maintenance of the race, but also who was boss. Effigies of male gods are noticeably absent.

Home and mother are written over every phase of the early, peaceful, horticultural societies. Before paternity was understood, beliefs linking motherhood with superior intelligence, reasoning power, and magical knowledge, apparently made men superfluous. Their sex role unknown, men were not even allowed to participate in religious services—except as the helpers of the priestesses, who were the first mathematicians. The Sanskrit word *matra* like the Greek word *meter* meant both "mother" and "measurement"—hence by derivation from "mother" came the words metric, mark, and mentality. And mathematics originally meant "mother wisdom."

In India, in early times, men believed that if only they could master the mathematical skills of the women, they too would be able to give birth. Imagine two of these guys meeting on the street. Says the first, "Well fella, how's it goin'?" To which the second replies, "Not too good! Them female figures are givin' me the fits! I got a headache, an' I ain't matriculated for months!"

The power of the motherhood idea was sustained by the close connection between the birth-giving female and birth-giving Mother Earth. Thousands of feminine names have been given Earth and its parts, for example the names of many nations: Libya, Russia, Anatolia, China, Scotia (Scotland), Eryou (Ireland). In her many guises Gaia, the Earth goddess, combined images of home and mother. Most of this primeval insight was lost when agriculture, another (more problematic) invention of women, produced the surplus grain harvests that made transportable food a bargaining tool and a weapon, the basis for patriarchy and the building of cities, a justification for standing armies and eternal wars—because when, as in the USA today, a billion dollars a day are being spent on armed forces you just can't have them standing around doing nothing.

Revealed religions of the Book, founded on agricultural ideas rather than on those of gatherers and hunters, pointed societies away from Earth, away from mother, toward more abstract realms.

In its search for power and control, modern science has inadvertently rediscovered many of the ancient truths regarding the human-Earth relationship. For example, each one of us is composed of Earth's *surface* materials—a fact that calls most mining into question. Our body/minds are shaped physically and psychologically by immersion in the thin green skin of the planet. Sandwiched like bacon-bits between the air above and the land beneath, we are deep-air animals, related not only to each other but to all the different kinds of existing organisms. Human languages—the vehicles of thought, of the arts and sciences—are derived from the physical experiences of living in place. We are Earthlings, and Mother Earth is again getting at least equal billing with Heaven and Hell.

Fortunate are we ex-urbanites who live surrounded by Nature's miracles. How much easier it is here—rather than entombed in cities—to feel the wonder of the living Earth, sensing with awe its mysterious harmonies. Here, in blossoming May, the dream of Mother Earth restored to her full diversity and grandeur, with our prodigal species once again a co-operative, responsible, ethical member, still seems possible.

Hail to all mothers and to their antique Earth-based beliefs, much more ecological, wise and reasonable than those that now rule Washington, Ottawa, and Victoria. May men once again take up the study of Mother Wisdom. Who knows, this time they may matriculate.

SERMONS IN STONES

One of the most important biological/ecological ideas of the last century is the insight that the world of everyday experience presents itself to us as organized in series of wholes and parts. Earth is a whole planet, yet also a part of the solar system which is part of the Milky Way Galaxy, in turn part of the cosmos. In the other direction, the whole Earth consists of large parts we call lithosphere, atmosphere, hydrosphere and biosphere, each further decomposable to constituents such as stones, oxygen, water, and organisms. All these are composed, smaller and smaller, of molecules, atoms, sub-atomic particles and energy waves.

Arthur Koestler invented a useful terminology, calling each level of organization a "holon" and any series of related holons a "holarchy." He noted that holons are two-faced, looking in opposite directions like the demigod Janus usually depicted with one laughing face and one crying face. Upward and outward in any holarchy each holon is a part of the holons above; downward and inward each holon is a whole to the holons it comprises. So every object, every holon, is simultaneously a *whole* to the parts of which it is made and a *part* of that which surrounds it. In this sense, all material reality is composed of holons.

Although Koestler did not further explore the implications of his Janus analogy, it seems likely that the happy face of each holon is its ecological face, looking upward and outward. The physiological face, looking downward and inward, is the sad one. Perhaps it weeps in recognition of the fact that it has become the tool of reductionist science, a major cause of Earth's current woes.

A stretch of beach by lake or sea is a holon, and a speckled granite stone nestled in the sand signifies its source, eons ago, as continental crust floating on the heavier magma that underlies the oceans. This stone-holon is composed of several kinds of holon crystals—feldspars, quartz, and micas—symmetrically built from molecules and atoms of sodium, potassium, aluminum, iron, silicon, and oxygen. The last two elements joined make hard, durable silica or quartz. As granite weathers away, losing its softer feldspars and micas, the silica crystals remain as sand. Tiny quartz grains on the strand may be a billion years old or older since initial crystallization. Time is measured in the hourglass by flows of the timeless sands.

When dissolved and separated, the constituent holons of a silica sand grain—oxygen and silicon—can be taken up as fundamental constituents of organic holarchies. On land the sand particles yield silicon to the roots and rhizomes of horse/tails (*Equi/setum* in Latin), helping to form the skeleton of these ancient plants and explaining their abrasive second name, scouring rushes. In water bodies the slowly dissolving quartz provides silicon for the shells and spicules of small animals and plants such as diatoms whose tiny opaline cases rain to the bottom, forming layers of sediment known as "diatomacous earth." Given time, pressure, and heat, such siliceous deposits at sea bottom may be formed into hard-rock quartzite, later elevated as parts of mountains, weathered once more into sand, then water-washed again to a beach for another long cycle involving both the organic and the inorganic.

A sand grain is a marvellous little holon, not only a potential contributor of its holon parts to many forms, both organic and inorganic, but itself a former part of a stone, a boulder, a rock formation, part of Earth's crust and mantle, part of Earth within its solar system, all congealed star dust from the explosion of a supernova four or five billion years ago. Set in context, it brings to mind the relatedness of all Earth's parts, of things incorporated in other things through long slow cycles, of creativity and beauty endlessly expressed.

"To see the world in a grain of sand" is to see the grain of sand as a holon in its Earth context, a reminder that under/standing (attending to the holon under-parts) ought always to be balanced by per/ceiving (attending equally to the enveloping holons). In/sight at its best can give only half the truth; out/look gives the other half or sometimes more. Searching within frequently answers the question "How?" Searching without provides clues to the more fundamental "Why?"

The tendency of science in a power-seeking society such as ours is to neglect the per/ception of context (the ecological approach) and to accentuate in/spection (the physiological approach). Reductionism is favoured by scientists as the quickest and easiest path to "know-how," to gaining control of things in order to manage them for human purposes. Medicine provides many examples, such as the search for cures of cancer and heart disease within the human holon rather than attempting an outer clean-up of its poisoned context: the industrialized environment of polluted air, water, soil, and food. The interior "magic bullet" approach will work temporarily, but only attention to the solution of outer problems will have lasting effects.

That the sad face of Janus is reductionist, inward-looking, away from seeing the Earth whole, seems a sound assumption. The happy, optimistic face that looks upward and outward, mirrors hope that the worth of Earth—a very large and miraculous stone in space—will universally be recognized in the near future.

Here endeth the sermon.

HAIKU: THE HUMAN IN NATURE

Temple bells die out
The fragrant blossoms remain sinking
A perfect evening

—Basho (1644–1694)

The names of the dead
deeper and deeper
into the red leaves

—Buson (1716–1784)

First autumn morning:
the mirror i stare into
shows my father's face.

—Murakami, Kijo (1865–1938)

The Haiku poetry of Japan is rooted in pre-Buddhist Nature-worship, conveying sentiments that today resonate with a sensible Earth-based ethic. These short impressionistic verses are "primitive," not in the sense of inferior but in the original meaning of "first." Primitive religious beliefs, judged by their consequences for sustainable living on Earth today, are more realistic than the abstract other-worldly faiths that have largely displaced them.

When we find our way past three thousand or more years of the Abrahamic religions whose transcendent God has been endowed by the prophets with an immoderate interest in human selves, especially the male ones, we rediscover the perennial wisdom that alone is ecologically reasonable: the Earth is sacred. This immanent Nature in which we are immersed is our evolutionary source. Within it we live, move and have our being. And to Earth, at the end, each of us returns.

Relieved of the burden of un-Earthly spirituality that renounces common sense, we can begin to see ourselves as we truly are: Earthlings, children of the life-giving, life-sustaining, blue-and-white planet.

This unifying idea carries important ethical implications. Believers understand that they are related parts of a greater organic/inorganic whole. Earth, in whose skin they are enveloped, is recognized as the prime reality that merits their wonder and devotion. Morality and ethical behavior are extended beyond the human race, which is perceived to be an important part of Earth, though not the single all-important part.

The philosopher Spinoza conceived Nature as the unity of all-that-is, identifying it with God. The belief that "God is all and all is God" has been called pantheism, a conception of great antiquity. According to Spinoza, the Nature that we know is one inclusive unlimited substance with many attributes, though revealed to people in only two ways: as material things (the outer view) and as mental ideas (the inner view). The way we live in the world, and the way we think about it, are all of a piece.

As our bodies and minds are inseparable, so our outer perceptions and our inner conceptions are inseparable. The perceptions of Semitic tribes wandering in austere deserts of the Middle East invited concepts of a transcendent sky-god always with them, promising a more secure life in the future. In contrast, the percepts of the ancestral Japanese who lived in a pleasant and bountiful environment suggested a harmonious relationship with sacred Earth, whose natural manifestations, called "kami," were worshipped at thousands of shrines: trees, caves, waterfalls, lakes, mountains. Thus the Japanese inherited a tradition of gratitude and respect for life, a deep appreciation of the beauty and power of Nature, a preference for purity, for the simple and the unadorned.

Integral to Japanese aesthetics is the idea of *Wabi-Sabi*, translated as elegant simplicity, rustic yet refined. This was the basic philosophy of such famous writers of Haiku as Basho who considered the immersion of self in the impersonal, ego-less life of Nature as essential to poetic creation. Basho was confident that Haiku, as a serious art form, could point toward an invaluable way of living. Like other classic poets, he expressed a faith in the goodness of Earth and its superb phenomena as purifiers of the human mind/spirit.

Can we purify our spirit/minds by expressing in simple verse the deep bond between ourselves and Nature? In a meditative poetic sense, it seems a worthwhile exercise.

For the writing of Haiku verse, several structural rules are commonly accepted:

1. The form consists of 17 syllables in three lines, usually 5-7-5.
2. Two separate thoughts are expressed—a short "fragment" which is either the first line or the last, and a longer "phrase" either the last two lines or the first two—each enriching the other imaginatively.

Composition reflects the philosophy and aims of the poet. To express the this-world orientation, the following rules are suggested:

1. Subjects are taken from Nature and from daily life, conveying vivid impressions.
2. The focus is outward on real world images, downplaying the personal.
3. A word or phrase suggesting the season of the year is usually included.

4. The present tense is used, dwelling in the here-and-now rather than in past or future.
5. Punctuation is minimized; ambiguity is acceptable.

So many people
Watching TV this fall night
Sad-faced the full moon

Quick night wind, warm rain
Sweet Balm-of-Gilead scent
March resurrection

Cone of dazzling snow
On each red bunch of berries
New Year's jewels

A bear on the path!
the philosopher runs like
everyone else

The point of Haiku
To be here now in Nature
Get out of your head

5. WHAT ON EARTH IS LIFE?

WHAT ON EARTH IS LIFE?

Although definitions of the meaning of "life" are various, none recognizes the importance of context. A partial explanation of the attribution of "life" only to organisms is the fact that the Ecosphere—the context and source of organisms—has until recently been invisible. Further, the metaphorical nature of abstract language has conflated "organisms" and "life." But just as the living parts of an organism depend on the vitality of the whole, so living organisms depend on the energetics of Planet Earth from which they evolved and by which they are maintained. From an ecological viewpoint Planet Earth, the inclusive supra-organic Ecosphere, is the best and most logical metaphor for "life" in its largest sense.

Introduction

Countless books and articles have been written on "life" and its meaning. The justification for yet another exposition is three-fold. (1) The question of what constitutes a valid *definition* of life needs clarification in the interests of better understanding between scientists and non-scientists. (2) Now that Earth is accepted as a functional system (Schneider 2001) with a "geophysiology" (Lovelock 1987), the ancient idea of extending "life" from organisms to Earth invites further evaluation. (3) The metaphorical nature of language, directing attention to phenomena deemed alive or dead, shapes cultural values and behavior patterns in important ways. When "organism" is taken to be the only correct metaphor for "life," organisms assume central importance in human affairs. Were "Earth" accepted as an alternative metaphor, significant changes in the human-Earth relationship would follow.

Because human languages are based on the physical experiences of existence on Earth, abstract qualitative concepts are often expressed or pictured as familiar things. Hence the natural tendency to confer thingness on "life," to equate it with organic bodies capable of certain performances such as metabolizing and reproducing—an example of Whitehead's "fallacy of misplaced concreteness." Morison (1971) pointed out the circularity whereby ideas of "livingness" are obtained from observing living things and of "deadness" from observing dead things, then turning the descriptions into nouns—"life" and "death"—that endow them with attributes of concrete form. Organisms express "livingness," from which has come the usual definition, "Life = Organisms."

Three Kinds of Definitions of Life

A compendium of definitions of life, with thoughtful comments, is given by Applewhite (1991). They range from philosopher Herbert Spencer's "Life is the continuous adjustment of internal to external relations," to scientist Ernest Borek's "Life may be defined as a system of integrated co-operating enzyme reactions," indicating that philosopher and scientist differ as to what constitutes an adequate definition. Thus the answer to "What is Life?" hinges on a prior question: "What constitutes an acceptable definition?"

Hughes (1998) identified three kinds of definitions that frequently overlap:

1. the reportive, or vernacular, as given in dictionaries;
2. the stipulative, of which one form is the operational, favoured by scientists setting up testable hypotheses;
3. the essentialist, favoured by philosophers attempting to get to the core of the matter.

Confusion is likely whenever definitions of "life" are presented without prior "definition of the definition" being used.

Reportive definitions reflect the standard usage. A typical dictionary definition of life: "that property of plants and animals which makes it possible for them to take in food, get energy from it, grow, adapt, . . . reproduce; . . . it is the quality that distinguishes a living animal and plant from inorganic matter or a dead organism." Describing the *quality* of "life" as the difference between the living and the dead is good enough for everyday experiences in which people intuitively, though perhaps incorrectly, make culturally appropriate responses to objects perceived as animate or inanimate.

Stipulative definitions fix particular meanings on words. One type, the operational definition, appeals to scientists. It takes the form: "X is alive when it exhibits measurable attributes a, b, c . . ." or, more dogmatically, "Life is measurable attributes a, b, and c. . . ." Here the meaning of "life" must necessarily fit the procedures used to measure it. Applewhite noted that "each biological specialty tends to describe the life process it studies in terms of a particular apparatus or technique of investigation," which is what one would expect of those collecting data at the molecular level, the cell level, or the organism level. Of the latter an example is the proposal of Chao (2000) that organic evolution by natural selection is the best operational definition of life, and that it should be the criterion in future searches for extraterrestrial life. Thus his definition stipulates that life can be recognized as present, here or elsewhere in the universe, if a specified operation—the ingenious "serially transferred labelled release experiment"—yields measurable positive results.

Essentialist definitions appeal to the philosophically minded, both scientist and non-scientist. They are attempts to grapple with important concepts in ways that are more inclusive than the vernacular and the quantitative/measurable. In the words of the logician, "The correctness of an essentialist definition cannot be determined merely by an appeal to standard usage like a reportive definition, nor by an appeal to its usefulness like a stipulative (operational) definition. Essentialist definitions really need to be understood as compressed theories; they attempt to express in succinct form a theory about the nature of what is being defined. Thus, assessing an essentialist definition involves assessing a theory, and this goes far beyond questions about the meaning of words" (Hughes 1998).

A famous essentialist definition is the equating of life with "negative entropy" by Schrodinger (1944). Another, more recent, is Capra's (1996) three-part definition of "living systems" as pattern (autopoiesis), structure (dissipative), and process (cognition), thus synthesizing ideas of Maturana and Varela, Prigogine, and Bateson. Sometimes the essentialist and the operational definitions are combined as by Chaisson: "Living systems stay alive by steadily maintaining themselves far from equilibrium . . . in fact, unachieved equilibrium can be taken as an essential premise, even an operational definition, of all life" (Applewhite 1991).

All Definitions are Theory-Based

Reportive and essentialist definitions tend to irritate those for whom the measure of truth is supportive data. They sympathize with Lancelot Hogben's exasperated "Biology is not the science of 'life' [because] science is not about the study of abstract nouns," and Ernst Mayr's statement that attempts to define life are futile because "there is no special substance, object or force that can be identified with life" (Applewhite 1991). If it cannot be measured, it is meaningless. Just here is the sticking point between "the two cultures," the scientific and the literary-scholarly.

Science champions a particular method of perception, and from long training scientists are prone to equate reality with the material/measurable. For example, the book *Consilience* by E.O. Wilson (1998) is entirely devoted to support of this thesis, extended to the arts. From equally long philosophical training the literary-scholarly are assured that diagnoses such as Wilson's are not the whole story. For them, immaterial things such as mind, religion, and culture are not identifiable simply as symptoms or epiphenomena of the material world. When scientists assume that their operational definitions of life are essentialist, as close to reality as one can get, embracing all that is worth notice, the philosopher's hackles rise and with reason. After all, they argue, experience reveals a reality with both quantitative and qualitative aspects, and the latter merit as much serious attention as the former. Further, they say, the scientist's operational approach defines life away rather than explaining it.

Musing on the penchant of scientists to fix the meaning of things by mathematical descriptions of physical quantities, Applewhite (1991) voiced the fear that the operational meaning of life may move into dictionaries as its reportive (vernacular) meaning. Should the measurable become the common sense definition, then perhaps self-reproducing technologies will qualify as living. "I am haunted by an unwelcome premonition . . . Does all this herald the advent—and possible supersession—of what we know as natural life by artificial life?" The answer may be affirmative if essentialist definitions of life are stigmatized as unscientific and, in a science-based society, thereby judged worthless.

Critics of non-stipulative definitions do have a point. Reportive definitions of life tend to be tautological while essentialist definitions invite condensed theories that to the tidy-minded verge on speculative woolliness. Because their understanding requires book-length exposition, essentialist definitions miss the mark of easy comprehension that characterizes vernacular and operational definitions. On the face of it, just what is meant by Buckminster Fuller's (1975) essentialist, "Life is the eternal present in the temporal"?

Such questions aside, the essentialist idea—that definitions embody theories as to what is real and important—can be extended to the other two types of definitions as well. Thus the above-noted operational definition of life by Chao (2000) carries the theoretical presupposition, shared also by reportive and essentialist definitions, that life is a possession of organisms, and that in fact the two terms are identical: Life = Organism, and Organism = Life. "Is there life in the universe?"

means "Are there evolving organisms like us, or at least like our cells or organelles, somewhere out there?" Here is the foundational axiom, rarely if ever questioned, from which probes are launched in search of terrestrial and extraterrestrial "life."

Considering "Life" in Context

An alternative to the view that organisms possess "life" is that "life" possesses organisms. By this hypothesis the secret of "life" is to be sought outwardly and ecologically, rather than (or as well as) inwardly and physiologically. Insofar as much of science is reductionist, with a tendency to anatomize in its search for mechanisms, such an outward reorientation may seem unreasonable. Past practice has been to seek the secret of life lower and lower in the hierarchy of organization that descends from organism to cell to organelle to DNA and RNA, to peptides and left-handed amino acids.

Whatever life may be, it necessarily involves ecological relationships. DNA is conceived as alive when functioning within living cells. As with viruses, the matrix somehow contributes to or confers vitality. Similarly the identification of living cells in the tissue of an organ implies the aliveness of the latter. Again, an organ such as the heart continues its animated beating as long as it is intimately related to the living organism that envelops it. In these three examples of "living things"—DNA, cell, organ—the context is accepted as alive, an essential life-giving system and not just a life-support system. Extending the logic, organisms are alive thanks to their embeddedness in larger encompassing "living" systems: the sectoral geographic ecosystems that Earth comprises. Note that "ecosystem," as here used, is a volumetric entity, an inclusive piece of Earth-space, and not a synonym for communities plus their vague surroundings.

The idea is implicit in the definition of "ecosystem" by Tansley (1935) who originated the term. "Though the organisms may claim our primary interest, when we are trying to think fundamentally we cannot separate them from their special environment with which they form one physical system." Tansley was a botanist and his concept of ecosystem was an extension of his interests in understanding plant communities. Nevertheless the logic of his fundamental idea, that organisms cannot be separated from their physical milieus, locates "life" more reasonably in Earth's inorganic-organic geoecosystems (Rowe and Barnes 1994, Rowe 1997) than in organisms *per se*.

A similar interpretation can be drawn from the work of Lindeman (1942), one of the first ecologists to make the ecosystem concept central in his trophic-dynamic studies of lakes. "The discrimination between living organisms as parts of the biotic community, and dead organisms and inorganic nutrients as parts of 'environment,' seems arbitrary and unnatural . . . this constant organic/inorganic cycle of nutritive substance is so completely integrated that to consider even such a unit as a lake primarily as a biotic community appears to force a 'biological' emphasis upon a more basic functional organization." Lindeman's "basic functional organization" was the lake ecosystem, a better candidate as the purveyor of "life" than the organic parts alone.

Although Tansley named ecosystems the basic units of nature on the face of the Earth, and Lindeman contrasted them with the biotic community as the more basic functional organization, neither took the further step of attributing "life" to these supra-organic, place-specific, air/land/water/organism ecosystems that in sum are displayed as Earth. With the exception of Lovelock and his followers, the same hesitation is evident throughout the scientific literature of the last century: nevertheless scientists have occasionally skirted the idea that Earth/environment promotes "life;" that each organism is embedded in a "system-of-life," and that it takes the entire biosphere (more broadly, Earth as Ecosphere) to make the concept "life" meaningful (Davies 1988).

Is Earth Dead, but for Organisms?

Biologists generally agree that living organisms evolved from Earth, are made from Earth, are sustained by Earth, and go back to Earth in timely cycles. Nevertheless, resistance is strong against the idea that Earth and the sectoral geographic ecosystems it comprises are in any sense "alive." Philosophic scientists usually go no farther than naming Earth "a life-support system." Life is denied planet Earth despite the fact that no organisms known to science could ever have existed and evolved without the Earth/ecosystem context, and despite the fact that organisms separated from Earth's ecosystems cease to exist. Imagination as to what may be "alive" stalls at the organism level. If X is not an organism, X is dead.

Even Lovelock (1988) accepted this logic and attributed life to Earth only because of bacteria and other primal organisms that colonized the pre-Gaian surface. In his thesis the planet was dead before organisms appeared. Gaia was non-existent until gifted with "life" by bacteria that (apparently without Earth's vital assistance) miraculously appeared in the early sea-soup. Lovelock's thesis, ascribing the source of Earth's animation to organisms, forced his conclusion that Gaia must itself be an organism, a super-organism. In effect the organic parts, though inconceivable without the whole, must somehow have conferred life on the whole. This concept is a high-level variant of the reductionist faith, vigorously attacked by Chargaff (1997), that wholes are "nothing but" assemblies of parts and therefore that the secret of life will ultimately be discovered by molecular biologists. My argument is that Earth should not be confused with organisms. Earth is not a super-organism. Earth is supra-organic.

The same idea concerning the Ecosphere's sectors has been expressed by Rapport et al. (1991) in these words: "Ecosystems are, to be sure, a supra-organismic level of organization, but are not super-organisms since each level in the hierarchy has both unique properties found only at that level and parallel properties with other levels. Accordingly ecosystems are not organisms, but there are analogous properties that may or may not function in the same manner at the two levels."

Ecosphere and its Ecosystems Invisible to Insiders

The idea that organisms alone are alive and that any "life" Earth possesses is due to its organic constituents seems reasonable to inside-the-ecosystem viewers. Embedded in Earth as deep-air animals, humans could not (until recently) see their matrix as a functional unity. Four hundred years before satellite vision showed the Earth-whole, science established its common-sense perspective within what appeared to be a fragmented world. Physics and chemistry arose to study the commonalities of the fragments. Geology, hydrology, meteorology, and biology followed, examining and explaining particular species of the fragments: thus the human environment was long ago divided into the organic and the inorganic, the biotic and the abiotic, the animate and inanimate, the living and the dead. The organic/biotic was firmly established as synonymous with the living/animate.

Biologists have described the characteristics of living organisms as including metabolism, development, growth, reproduction, evolution. Given these criteria, Earth fails to qualify as an organism and, given the premise "Organism = Life," Earth is pronounced dead. But if the two concepts "life" and "organism" are disengaged, Earth qualifies as "alive" by a different route that transcends and melds both organic and inorganic. The status of Earth is not that of a super-organism drawing its vitality from organisms as Lovelock argued. Earth is supra-organic, a higher level of organization than the organic and inorganic. The reality of surpassing importance—the blue planet—with its integration of the organic and inorganic, expresses life as well as harmonizing it with death in perennial cycles.

Metaphorical Nature of Abstract Concepts

Much of verbal language is metaphorical, and changes in paradigms, theories, and concepts mirror changes in the metaphors by which they are expressed. Linguists such as Lakoff and Johnson (1980) suggest that the original meanings of words and sentences are based on the way humans experience their physical being in the world. Examples are the many word-variations that in their meanings incorporate "in" and "out," "up" and "down," "above" and "below," "standing" and "lying." By analogy and metaphorically the original meanings, derived from material existence, have been extended to more abstract words; for example, the mental phenomena "under/standing" and "super/stition" take their meaning from what "stands below" and what "stands above."

A commonplace of metaphorical language is the equating of things familiar and material with things abstract and qualitative, such as "heart" with "courage," "brain" with "mind," "organism" with "life." The concepts of each pair do not refer to the same phenomenon. But because they are linked, the material member makes a useful metaphor for the immaterial. The conventional metaphor for life is organism. A legitimate question is whether there might not be a better metaphor, meaning by "better" a material concept which, by more accurately reflecting the structures and processes of experienced reality, broadens, deepens, and tends to unify understanding and sensitivity. In short, can a metaphor be found that better harmonizes human thoughts and actions with the world as experienced?

An ecological answer is that "Earth" is the better metaphor for "life." From its creative "mega-symbiosis" of the inorganic and organic over vast evolutionary time came today's diversified world. So far as is known at present, animated creatures occur only on this planet in the context of its inorganic/organic matrix. Organisms *per se* are born out of Earth, are sustained by Earth, evolve in Earth's skin, and in death and dissolution return to it for cycles of rebirth.

So What?

The essentialist definition "Life = Earth" does not pretend to answer the question "What is Life?" It simply replaces a narrow physiological metaphor that accepts organisms as the carriers of life with the broader ecological metaphor identifying Earth as life's purveyor. Adoption of the latter would not negate such operational definitions as the one proposed by Dr. Chao. Detecting evolving organisms on Mars, or on the Jovian moons, would indicate that these celestial bodies, like Earth, also possess some degree of "aliveness" within the vital solar system. Such unlikely findings would still leave the question "What is Life?" unanswered.

Nor would the definition "Life = Earth" make the subject of biology any less important. It would widen the horizons of those attempting to explain or duplicate primeval "life," adding to the emphasis on the physiology of animate matter its equally vital ecology.

As the current interest in rescuing and preserving biodiversity is revealing, the understanding of "livingness" and its preservation is as much a matter of attention to Earth's ecological systems as it is to organisms. Organisms, species, genetic diversity, cannot be preserved in health without preserving the integrity of the geographic ecosystems of which they are dependent co-evolved components (Rowe 1998). The key question for the human species at the moment is how to manage itself in the interests of maintaining the evolutionary creativity and health of the Ecosphere's ecosystems on which the welfare of all depends.

Awareness of Earth as the giver and maintainer of life, shifting the focus from organisms to the larger system that is their mutual source and support, might in time revivify and re-enchant a world that science, for several hundred years, has assumed to be dead. Such a shift could redirect mainstream intellectual, emotional, and ethical activities toward the goals of harmonizing humans with their ecosystem homes, repairing the damage done to Earth and to living things in general because of past ignorance.

Were it widely adopted, the most consequential impact of the metaphor "Earth = Life" would be on the reportive or vernacular definition, influencing popular usage and understanding. By extending the all-important life-centre beyond organisms and *Homo sapiens sapiens* to the Ecosphere's creative, sustaining, enveloping matrix, the new metaphor would point away from the traditional anthropocentric-biocentric ethic whose unhealthy results are more and more evident worldwide. It would urge an ecocentric ethic in harmony with such realistic evolutionary/ecologic thoughts as, "In the beginning was the World," and, "first the Earth"—a cosmopolitan message that in these troubled times is neither inimical to a universal science nor to religion in its fundamental "binding together" sense.

Acknowledgements
For helpful comments in the preparation of this article I thank Drs. Burton V. Barnes, David Rapport, Glenn Albrecht, and Theodore Mosquin, the latter lending me his term "megasymbiosis" to describe the Ecosphere.

References

Applewhite, E.J., *Paradise Mislaid: Birth, Death, and the Human Predicament of Being Biological* (New York: St. Martin's Press, 1991) 8-13.

Capra, Fritjof, *The Web of Life* (New York and Toronto: Anchor Books Doubleday, 1996).

Chao, Lin, "The Meaning of Life" *BioScience* Vol. 50 No. 3 (2000) 245-250.

Chargaff, Erwin, "In Dispraise of Reductionism" *BioScience* Vol. 47 No. 11 (1997) 795-797.

Davies, Paul, *The Cosmic Blueprint*. (New York and Toronto: Simon & Schuster Inc., 1988).

Fuller, Buckminster, *Synergistics: Explorations in the Geometry of Thinking*. (New York: Macmillian, 1975).

Hughes, Wm., *Critical Thinking*, 2nd ed. (Peterborough, Ontario & Orchard Park, New York: Broadview Press Ltd., 1998).

Lakoff, George and Johnson, Mark, *Metaphors We Live By* (Chicago and London: The University of Chicago Press, 1980).

Lindeman, R.L., "The Trophic-Dynamic Aspect of Ecology" *Ecology* Vol. 23 (1942) 399-418.

Lovelock, James, *The Ages of Gaia* (New York: W.W. Norton & Company, 1988).

Lovelock, J.E., *The Geophysiology of Amazonia: Vegetation and Climate Interactions*. ed. R.E. Dickinson (New York: Wiley, 1987) 11-23.

Morison, Robert S., Death: Process or Event? *Science* No. 173 (1971) 694-698.

Rapport, D.J., H.A. Regier, and T.C. Hutchinson, "Ecosystem Behavior Under Stress" *The American Naturalist* No. 125 (1991) 617-640.

Rowe, J.S., and B.V. Barnes, "Geo-ecosystems and Bio-ecosystems" *Ecological Society of America Bulletin* No. 75 (1994) 40-41.

Rowe, J. S., "Defining the Ecosystem" *Ecological Society of America Bulletin* No. 78 (1997) 95-97.

Rowe, J. S. "Biodiversity at the Landscape Level" *The Living Dance: Policy and Practices for Biodiversity in Managed Forests*. eds. F.L. Bunnell and J.F. Johnson (Vancouver: UBC Press, 1998) 82-95.

Schneider, Stephen H., "A Goddess of Earth or the Imagination of a Man?" *Science* No. 291 (2001) 1906-1907.

Schrodinger, Erwin, *What is Life?: The Physical Aspect of the Living Cell*. (Cambridge: Cambridge University Press, 1944).

Tansley, A.G., "The Use and Abuse of Vegetational Concepts and Terms" *Ecology* No. 16: (1935) 284-307.

Wilson, E.O., *Consilience: the Unity of Knowledge*. (New York: Alfred A. Knopf, 1998).

❦ THE MECHANICAL AND THE ORGANIC

Reprinted with permission from *The Structurist*, No. 35/36, 1995–1996. University of Saskatchewan, Saskatoon, pp 13–19. Edited for this collection.

Writers of fiction looking to the past have imagined our ancestors as hairy cave men and women—strong in body but of little wit. Others, looking ahead, have projected a race strong in mind but almost disembodied—big brains on push-button fingers. Such is rationality's prestige that the latter vision beats the former, hands down. Who covets the healthy outdoor life of our nearest relatives, the carefree chimpanzees mindless of tomorrow? How much better the angst of cultural evolution by technology whereby brains are enlarged via the computer's microchips, while bodies pale and soften in the glow of the cathode tube!

Choice between chimps or chips, between body or mind, is not an option. We humans are body/mind, a necessary melding of the material and immaterial. Yet often we speak as if one is detachable from the other, with the implication that the more valuable *mind* can be disconnected from the less valuable *body*. Hope for such a separation springs eternal in the human breast, particularly when the ego confronts death. But, at least in *this* existence, the aspiration to divorce mind from body is perilous insofar as it depreciates the source of health which is the body and, by extension, the world that generated body/mind and sustains it for a lifetime.

My thesis is that the theorists of both Modernism and Postmodernism sacrifice body to mind, the real to the ideal. They put their faith in the artifacts and abstractions of intellect while devaluing their source and support: Planet Earth. The mechanical supplants the organic. Thus, virtual reality—faking Nature symbolically "on-line"—puts the future of Nature precariously on the line.

Mind versus Body

A philosophic tradition with roots beyond Plato supports the notion that ideas—the tissues of mind—are the ultimate realities compared to which phenomenal bodies are but shadows. "You must forgive me, dear friend," Socrates said to Phaedrus, "I'm a lover of learning, and trees and open country won't teach me anything."

The implications of this platonic theme for a high-tech society were essayed in 1928 by E.M. Forster. In his story "The Machine Stops," people are housed or "hived" in air-conditioned comfort below the surface of an Earth that has mostly been devastated. Anticipating the TV screen and the Internet, Forster describes the coming age of virtual reality where everyone converses electronically (never face to face), mostly about *ideas*. Vashti, a music specialist, observes that visiting Earth's surface is vulgar and faintly improper for spiritually minded people; it is contrary to the spirit of the age, because air and stars and mountains "give no ideas." A popular slogan is, "Beware of first-hand ideas!" Those who want to know what Earth is like can listen to lectures on it, compiled from lectures formerly given on the basis of even earlier lectures.

Vashti's rebellious son, Kuno, discovers his long-disused muscles and concludes that his body is the measure for all that is lovable, desirable and strong. He climbs a shaft to the Earth's surface but is soon discovered and dragged back below. He says to Vashti:

> Cannot you see that we are dying and the only thing that really lives down here is the Machine? It has robbed us of the sense of space and of the sense of touch, it has blurred every human relation and narrowed down love to a carnal act, it has paralyzed our bodies and our wills, and it compels us to worship it. We only exist as the blood corpuscles that course through its arteries . . . I have only one remedy, to tell men again and again that I have seen the hills of Wessex . . . the dear ferns and the living hills.

Three quarters of a century after it was written, the story has a prophetic ring. An anti-naturalistic preoccupation with ideas and ideals, fed by a technology of symbols, lies at the root of the West's denigration of body as compared to mind, along with its negligent treatment of Earth. A chief function of today's ecofeminism is to remind all genders that bodies exist as expressions of natural rhythms—living, breathing, reproducing, dying—tuned to Sun, Moon and Earth, rather than to masculine philosophies. Perhaps today the chief function of art and aesthetics is to teach again the importance of the sensory component of perception, "to bring us to our senses," exposing idealism's bloodless logic that dismisses as secondary the intuited, richly sensuous, organic world.

The World of Spontaneous Experience

David Hume, the Scot skeptic, demolished both mind and matter. All we know of either, he said, are insubstantial "impressions"—thus the witticism, "No matter, never mind." Most of us ignore the sceptic's logic and accept as fact that we live our daily lives in a world of material things, even though "this universal and primary concern of all men is soon destroyed by the slightest philosophy." Rationality may suggest that a chair is nothing but an impression, a mind-image, or a constellation of sub-atomic particles whirling in mostly empty space, but still we perceive a chair as a back attached to a seat and use it as a four-legged support when sitting down to an apparently material breakfast. Hume was fully aware of this disparity between logic and the reality that Naess (1993/94) has called "spontaneous experience." It seems evident, wrote Hume, that "men are carried, by a natural instinct or pre-possession, to repose faith in their senses: and that, without any reasoning, or even almost before the use of reason, we always suppose an external universe" (Hume cited in Hargrove 1989). In short, the "natural instinct" that supports belief in a physical existence, with its spontaneous experience of valued material objects, is a body-wisdom that makes more sense than the philosopher's logic.

How strange that arguments have to be made for the tangibility of an experienced world of water and clouds, forests and flowers, humus and humans, while the "real" is attributed to intangibles: eternal forms, spirits, souls, symbols, words, language. Some today believe that "discourse" (literally, mental running to and fro) *makes* reality: an academic bit of froth that floats on an ocean of science/technology whose material artifacts provide the substantial matrix of urban living. In support of the experienced common sense world, Hargrove (1989) added two additional thoughts. Ordinary language is based most fundamentally on belief in physical objects, reference to which is largely what holds a language together and provides the basis for translation between languages. Further, aesthetics as a field of philosophy began with discussion of the qualities of natural objects, their beauty and sublimity. It would be a poor aesthetics that excluded the World of Nature and dealt only with the elegance of theories, the beauty of ideas and the charm of texts.

Skepticism about the existence of a natural world of inherent value, in which we are enveloped, seems less a sophisticated than a head-in-the-sand stance. But technology, along with an ideology preoccupied with our own species, blinds us to ecological perceptions.

Technology/Philosophy and Humanism

The technologies that people adopt influence their philosophic views. A classic example is invention of clockwork machines in the early Middle Ages, soon followed by the philosophic invention of a clockwork universe—which has served science well but Earth poorly. Ferguson (1966) championed the thesis that *concept follows conduct*: in adopting any appealing technology, men act and then they think. Ideas and sentiments arise out of repetitive physical acts. Men sailed before they made a cult of the sea; they killed with guns before they made a cult of the gun. Both cults are perpetuated in manly ego-boosting rituals played out again and again in sagas of the sea and in Western movies, where women are inconspicuous and never interfere. The technology of forearms, the gun, *makes* the Western myth; hence it would be ridiculous to negotiate a peaceful ending rather than shooting it out.

Some technologies are benign, and Ferguson suggested that those fostering non-combative interactions such as trade and commerce (!), with exchanges of information and communication between individuals and nations, should result in a cult of cooperation. Will global trade make for peace? Perhaps to the same extent as international football games and the Olympics.

Another optimistic postulate is that the way a technology is *assembled*, let alone the way it is *used*, may change society for the better. Thus the present generation of children, growing up with computers, will think ecologically because the computer is a technologic analogue of an ecosystem. According to techno-philosophers, the dramatic changes effected in the whole configuration of a spreadsheet by one keystroke on the PC will impress on youngsters the fact that exterminating a single keystone species can reverberate in a damaging way through an entire biotic community. Thus new technologies will inspire in society the same integrated "system-ideas" that they embody. Dream on!

School teachers, less confident than the technocrats, report a "dumbing down" of children brought up by TV *in loco parentis*. Without daily experience of the lively interplay of language between real people, a child's foundation for intellectual development is weakened. Some suspect that technologies may even change body/minds at the cellular level, "rearranging the brain's neurons to suit a TV-driven seven-second attention span, say, or an escalating desire for cinematic explosions" (Kroker 1996). This is a version of the yet-to-be-discredited proposition that given sufficient exposure, TV can turn brains into porridge.

Doubtless technologies mold attitudes, the medium generating its inherent message as Marshall McLuhan foresaw and as Jerry Mander (1991) and Kirkpatrick Sale (1995) forcefully argue today. Automobiles will not conduce to compact cities nor will weapons encourage peaceful thoughts. Machines made to serve efficiency will not foster a high valuation of leisure in the workers they displace. The Internet will not inspire close flesh-and-blood communities.

As technology influences attitudes so attitudes affect technology. The two co-evolve, and the configuration at any given time expresses deep historical-cultural forces. One major change-generating dynamic is the ancient belief-system combining faith in progress with faith that the human species is especially favoured by God. This guiding paradigm creates the climate in which rationality discovers its human-centred goals, dresses them in supportive ideas and sentiments, and pursues them with an evolving science/technology.

Narrow (Chauvinistic) Humanism

The deep dynamic that drives the attitudes/technology twins has various names: speciesism, anthropocentrism, homocentrism, even humanism in its arrogant sense. Humanism, as an expression of care for our species and of admiration for its positive accomplishments, certainly merits applause. Its dangers are claims to exclusivity, to selfishly valuing human welfare above all else on Earth. Variations on the chauvinistic, taken-for-granted, people-only theme comprise the literature of 99 per cent of the world's libraries, unconsciously expressing the narrow perspective from which humanity has viewed itself and the universe, especially in the last three or four centuries of modernistic thought.

Criticizing narrow humanism, Claude Levi-Strauss championed the humility of tribal people for whom "a well-ordered harmony does not begin with the self, but rather places the world before life, life before man, the respect for other beings before the love of self." (Emphasis added to Levi-Strauss's order of importance: world, life, man, other beings, self). This, he insisted, is not misanthropy but rather a critique of the strutting and shameless arrogance that makes man the lord and master of creation. The rights that one can and should recognize for mankind are only a special case of those rights that must be granted to the creative force of life. Care about mankind without simultaneous solidarity-like caring for all other forms of life, he said, leads mankind to self-oppression and self-exploitation (cited by Zimmerman 1994). Will care for Earth weaken the fragile bonds of democratic society?

Those firmly wedded to the *status quo* often express fear, genuine or feigned, that placing high value on Nature will necessarily devalue people, leading straight to totalitarianism. Levi-Strauss's argument is directly opposite. A sole obsession with humanity, he argued, is the road to disaster. Grounding ethics within the segregated ghetto of Humanism will not protect against such evils as racism and fascism. Consciously reviewing the meaning of words and identifying ideologies in order to control them, as recommended by Saul (1995), is a call for virtues with no external referent. And no matter what virtues humanity ascribes to itself, some group will always assert its superiority by claiming to possess greater virtues and more "rights" than others. The anchor-point for a "principled humility" must be found in a greater-than-human reality, one that transcends every culture and every ingrown morality.

Along the same lines, Berman (1984) agreed with Wilhelm Reich that industrial democracy is dry tinder for the irrational, for fascism, precisely because it is so sterile, so "out of this world" in the sense that mind (disembodied consciousness) is at war with Nature. Nations and their political economies are rootless, for rootedness must spring from ecological awareness of what it means to live in place "where Nature is the model for culture because the mind has been nourished and weaned on Nature." "It is my guess," wrote Berman, "that preservation of this planet may be the best guideline for *all* our politics . . . the health of the planet, if it can be successfully defended against the continuing momentum of industrial socialism and capitalism, may thus be the ultimate safety valve in the emergence of the new consciousness." Here is the call for a post-humanist faith that, acknowledging the supra-human value of the Earthly context formerly ignored or slighted, corrects an ancient error (Gray 1995).

An Ecological Perspective: Post-Humanism.

The criteria of a "good" scientific hypothesis or theory are (1) its explanatory power, (2) its capacity to integrate disparate observations, and (3) its heuristic abilities that open new perspectives. These are the characteristics that, for example, confer superiority on the theory of evolution over the theory of creation. By the same criteria the best myth/theory of our time tells the story of Earth's star-dust beginning and subsequent elaboration mirrored in its interactive parts: improbable atmosphere, sea water, continents, soils and sediments, all evolving along with a spectacular complex of organic forms including *Homo sapiens*, one of today's twenty-five million vivacious species.

Scientists are prone to use the terms "life" and "organism" as if they are one and the same, reflecting the reductionist idea that the source of the energizing life-spark of "things like us" is strictly inside our skins. The question, "When did life begin?" is assumed to be the same as, "When did identifiable organisms appear?" From an ecological viewpoint—i.e. granting importance to context—the question becomes, "at what stage in the history of Earth's increasing complexity did it engender identifiable organisms?" The spotlight shifts to planet Earth as the central marvel, the source and support of "life"—whatever that mystery may be. Bereft of Earth's permanent vitality, organisms lose their transient vitality. Neglect of Earth's health is a threat to all its constituents, organic and inorganic, now and in the future.

Note that the ecological perspective magnifies the value of what has traditionally been portrayed as a mostly dead environment. Liveliness exists in all things—in air, water, soil, as well as in organisms. All parts of the living Earth or Ecosphere are important, the inorganic equally with the organic. Earth itself is not an organism in the biologist's sense, nor is it a super-organism. The Ecosphere is *supra*-organismic—a higher level of integration than the inorganic and organic parts it comprises. A logical basis is provided for a drastic move in philosophic position from homocentrism to ecocentrism.

The viewpoint is post-humanist, radically ecological, favouring a focus broadened to include the matrix enveloping humanity. It acknowledges, with a celebratory sense of awe and wonder, the source of creativity in undomesticated Earth. Although it affirms that the protection of Earth-life requires a global scaling down of the human enterprise, it is not anti-human. It recognizes the importance, within the human community, of an "emancipatory agenda" that furthers and supports democracy, equity, justice, and freedom from oppression for the individual. These values cannot be fully achieved nor long preserved within a culture of chauvinistic humanism.

Modernism, Postmodernism and Virtual Reality

The enthronement of reason is the chief achievement of post-Enlightenment "Modernism" in its rebellion against centuries of religious dogma and uncritical acceptance of authority. Science is acclaimed as Modernism's finest fruit, carrying in its train ideas of perpetual progress: greater and greater material well-being through economic growth, fuelled by science/technology. Congruent with science's search for unifying theories is Modernism's "discourse of universals," meaning belief in over-arching social truths, belief in all-encompassing ideologies, in single "best" political systems, in understanding human groups (e.g. the class, "women") by reference to biological or social "essences" (Ross 1988). Modernity's "emancipatory agenda" of democracy and human rights is a "universal" rooted in the ancient Judaeo-Christian theme of the importance of the individual before God, energized most recently by the French and American Revolutions that promoted the political and economic importance of freedom for the individual.

Postmodern theory arose as a challenge to Modernism's theses, though oddly excepting the axiomatic thesis of homocentrism. Disillusioned with the claims to hegemony of both capitalism and Marxist socialism, it rejects the "discourse of universals." It also rejects belief in essentialist categories, the search for "totalizing" grand theories, the demonology of single causes such as corporations are to blame, technology is to blame, people on welfare are to blame. It subscribes to some parts of Modernity's "emancipatory agenda," aiming to promote social justice among all social and cultural groups (Zimmerman 1994).

Much of the reaction against Modernism's "truths" is praiseworthy and defensible, but the French deconstructionists carried their rebellion much further. Symbols and language are for them the framers and carriers of reality. All explanatory narratives, all texts are power-motivated and must be so interpreted. A statement such as, "Scientists claim that over-population is a cause of deterioration in Earth's environment" is interpreted as a ploy by white middle-class men to elevate their prestige, while guilting the world's poor. Calls for world population control are veiled attempts to keep women subservient. No language of disinterest exists. All categories are cultural constructs, and "truth," for example, cannot be objective; in the final analysis it can only be an outcome of multicultural consensus.

Postmodernism's radical, subjective relativism marks the contemporary rebirth of idealism: *linguistic idealism*. In a world of signs and symbols, the physical—the material, the body, the Earth—is devalued and pushed into the background (Sessions 1995a). If, as earlier suggested, technologies invent or "find" their appropriate philosophies under the guiding hand of Humanism, then homocentric postmodernism was surely conjured into being by virtual reality. Both replace an objective world with subjective tokens and facsimiles. Reality is up for grabs.

When faith that Nature-as-Earth can be discovered is replaced by faith that Nature is socially constructed, as many "Natures" exist as ethnic groups and languages. Nature becomes just another "off-line" fabrication, a cultural projection that can be reinvented "on-line" in any way people want. Many academics explicitly accept this dangerous idea (for example, Berleant 1992 and Schama 1995) which makes the West Edmonton Mall as "wild" as Jasper National Park, Disneyland as charming as the Florida Everglades, and the dream-world of virtual reality better, cleaner, and easier to access than any Earthly reality—given only that the choices between better and worse reflect multicultural consensus. Further, the evolutionary biological-ecological basis of a specific *human nature* disappears into thin air. Human nature is whatever people want to make it by hooking ourselves to the appropriate technology.

Kroker (1992) suggested that contemporary people are "possessed" by virtual reality, dominated by the technotopian dream and its assumed inevitability ("You can't stop progress, so upgrade or die"), ruled more and more by the "will to virtuality." Their wish, he said, is to upload themselves into the great Internet in the sky, to shed the weak fleshly bodies for the hard perfection of wiring and machinery. Dreaming of disembodied existence, they want to become pure data. In the same vein Barlow (1996), foresees the wiring of human and artificial minds into one planetary soul, the realization of Teilhard de Chardin's noosphere—"sufficiently interesting to provide company for God." Barlow writes, "When the yearning for human flesh has come to an end, what will remain? Mind may continue, uploaded into the Internet, suspended in an ecology of voltage as ambitiously capable of self-sustenance as was that of its carbon-based forebears." The ultimate promise of the Internet is immortal mind, released from corpo-reality. Perhaps this is the last and most appealing option for a thoroughly postmodern, industrialized, technologized and urbanized species.

The Human Predicament is Fundamentally Ecological

In Forster's tale an electronic technology, separating person from person and isolating society from external world, promoted the elevation of "mind and ideas" over "body and earth." All facets of existence were wrapped in artifacts. Each individual was incarcerated in the prison house of technology, living in solitary electronic confinement. People communicated by a kind of Internet; they shrank from ecological contact with other persons, with other organisms, with Earth itself. Virtual reality replaced the biological-ecological realities of seas and continents, sun-warmed air, clean water, fertile soils, mountains and green forests. The fount of creativity and beauty, of physical and mental health, had been completely rejected in favour of a comfort machine whose only sustenance for the human mind was human artifacts, images, ideas. Is this a fair description of inner-city living today?

Ahead of his time, Forster identified the fundamental human problem as ecological, rather than sociological. His solution: be prepared to lay aside the technological garments that stand between the bonding of mind to vigorous body, and the bonding of body to life-giving planet. The first priority is the establishment of a healthy relationship between mind, body and Earth, which are one. In a context of ecological well-being (and here should be added, "within ecoregions") the way is prepared for solution of psychological and social problems. By accenting the importance of Earth as context, Forster implied that full human development necessitates connections with our organic evolutionary roots in Earth. Any technology that interferes, injuring body and mind, must be resisted. He speaks to us today as an ecologist, a neo-Luddite, a post-humanist.

Forster's contemporary dystopians (Zamiatin 1921, Huxley 1932, Orwell 1949) placed little importance on the Earth-body/mind relationship. Although Zamiatin came closest, the three did not identify the fundamental human peril as ecological but as sociological, i.e. as loss of liberty, oppression by an elite, control of the individual by "soft" technologies and, in the extreme, control by brute force. Their rallying cry was, "Citizens, don't trade your individual liberty for bread and circuses!" Thus they expressed the freedom-seeking goals of Modernism, as do the many contemporary varieties of "social ecologists" who claim to be radically ecological though their agendas betray them. They are not post-humanists, not ecocentric (Sessions 1995b). Their primary goal—estimable but too narrowly conceived—is *social justice* for one disadvantaged group or another. They co-opt ecological arguments to belabour the opposition, but their philosophy is thoroughly homocentric.

Postmodernism stands ready to carry Humanism to its ultimate limit. By rejecting the reality of a Nature inside and outside body/mind, by rejecting the evolved biological-ecological organic Earth, postmodern theorists assert that an objective world possessing intrinsic value is meaningless. Human nature is whatever humans want to make it. The human body is a failed project. Thenceforth evolution will mean a closer and closer linking of mind to machine, marrying the carbon-based brain to the silicon-based computer. The trend is clear and in the spirit of, "Join 'em if you can't lick 'em," women are called to be cyborgs, postmodern creatures melding animal, human and mechanical components (Haraway 1991). Symbol of the twenty-first century: the pregnant robot.

Faith that the universe is person-centred, faith that personal comfort and convenience are the aims of existence, faith that all new technology is good technology, will continue to call up a spectrum of solacing technological garments, prosthetic cocoons of which the most extreme obviate the need for an external world by projecting artifacts of it "on-line." An ersatz universe effaces all notions of its own unreality. Once embedded in it, surrounded by it, any thoughts of escape evaporate. By blanking out every intimation of deprival, virtual reality places the fate of body and Nature squarely on the line.

As the remaining fragments of the human "wild-within" and of nature's "wild-without" are invaded, caged or whittled away, postmodern's virtual reality moves in to take their places. Forster's closing lines from "The Machine Stops," phrased in the un-neutral gender language of his day, sound a vivid warning:

> Ere silence was completed their hearts were opened and they knew what had been important on Earth. Man, the flower of all flesh, the noblest of all creatures visible, who had once made god in his image, was dying, strangled in the garments that he had woven. Truly the garment had seemed heavenly at first, shot with the colours of culture, sewn with the threads of self-denial. And heavenly it had been so long as it was a garment and no more, so long as man could shed it at will and live by the essence that is his soul and the essence, equally divine, that is his body. The sin against the body—it was for that they wept in chief; the centuries of wrong against the muscles and nerves, and those five portals by which we can alone apprehend—glozing it over with talk of evolution, until the body was white pap, the home of ideas as colourless, last sloshy stirrings of a spirit that had grasped the stars.

References

Barlow, John Perry, Quoted by Mark Kingwell, "Geek with an Argument," *Saturday Night* Issue 3778 (1996) 75-77.

Berleant, Arnold. *The Aesthetics of Environment (*Philadelphia: Temple University Press, 1992).

Berman, Morris, *The Reenchantment of the World* (Toronto: Bantam Books, 1984).

Ferguson, Charles W., *The Male Attitude: What Makes American Men Think and Act as They Do?* (Toronto: Little, Brown and Company, 1996).

Forster, E.M., "The Machine Stops" *The Eternal Moment and Other Stories* (New York: Harcourt, Brace & World, 1928) 441-475.

Gray, John, "Back to Nature" *RealWORLD* Vol. 14 (1995) 4-6.

Haraway, Donna J., *Simians, Cyborgs and Women: the Reinvention of Nature* (New York: Routledge, 1991).

Hargrove, Eugene C., *Foundations of Environmental Ethics (*Englewood Cliffs, New Jersey: Prentice Hall, 1989) 45, 176-177.

Huxley, Aldous, *Brave New World (*Harmondsworth, Middlesex: Penguin Books Ltd, 1932).

Kroker, Arthur, *The Possessed Individual: Technology and the French Postmodern* (Montreal: New World Perspectives, Cultural Texts Series, 1992).

Kroker, Arthur, Quoted by Mark Kingwell, "Geek with an Argument," *Saturday Night* Issue 3778 (1996) 75-77.

Levi-Strauss, Claude, Cited by Michael E. Zimmerman, *Contesting Earth's Future: Radical Ecology and Postmodernity* (Berkeley: University of California Press, 1994) 92, 116-7.

Mander, Jerry, *In the Absence of the Sacred* (San Francisco: Sierra Club Books, 1991).

Naess, Arne, "Creativity and Gestalt Thinking" *The Structurist* Vol. 33/34 (Saskatoon: University of Saskatchewan, 1993/94) 51-52.

Orwell, George, *1984* (Hammondsworth, Middlesex: Penguin Boods Ltd, 1949).

Ross, Andrew, *Universal Abandon: the Politics of Postmodernism* (Minneapolis: University of Minnesota Press, 1988).

Sale, Kirkpatrick, *Rebels Against the Future* (New York and Don Mills: Addison-Wesley Publishing Company, 1995).

Saul, John Ralston, *The Unconscious Civilization* (Concord, Ontario: Anansi, 1995).

Schama, Simon, *Landscape and Memory* (Toronto: Random House of Canada, 1995).

Sessions, George, "Postmodernism and Environmental Justice" *The Trumpeter* Vol.12 No. 3 (1995) 150-154.

Sessions, George, "Political Correctness and Ecological Realities" *The Trumpeter* Vol. 12 No. 4 (1995) 191-196.

Zamiatin, Y., transl. by Mirra Ginsburg, *We* (New York: Bantam Books, 1921, 1972).

Zimmerman, Michael E., *Contesting Earth's Future: Radical Ecology and Postmodernity* (Berkeley: University of California Press, 1994).

EARTH AWARENESS: THE INTEGRATION OF ECOLOGICAL, AESTHETIC, AND ETHICAL CONSCIOUSNESS

Reprinted with permission from *The Structurist*, No. 41/42, 2001–2002. University of Saskatchewan, Saskatoon, pp 11–17. Edited for this collection.

Armed with the unifying theory of evolution, the disciplinary fields of cosmology, geology, and biology have provided a comprehensive story of how Earth and its inorganic/organic components came to be. The marvellous diversity and complexity of stars, planets, seas, lands, air, plants, animals, minds, and consciousness apparently emerged over vast time from the deceptive simplicity of astral matter/energy.

The how of it all is science's best story—a narrative of sheer wonder. Yet what has driven cosmic, stellar, planetary, and organismic development remains as inexplicable as why the heavens and Earth possess this dynamic creativeness. Answers to the what and why questions lie in the "mirage-realms of intuition and faith," beyond purely rational understanding—which perhaps is just as well. Contemplating the how narrative of evolutionary change over billions of years is sufficiently mind-boggling and rich in its ramifications.

This does not mean that I am another enthusiastic science-buff about to explain everything reductively in terms of biology, chemistry, and physics. While I decry sentences that begin, "As a scientist, I. . ." or "As an artist, I . . .", establishing impressive credentials usually bogus for what follows, I admit to having something of a science background in ecology, relating organisms to the world that envelops and includes them. Human ecology seeks to understand the many ways people are connected to the places where they dwell, their "geographies of hope," set in surrounding regions and on Earth in its entirety. Other synonyms for the whole Earth are "Ecosphere" and "Nature"—names for the reality with which we are directly in touch, from which all organisms including the human have come, to which they belong and to which they return.

To locate intuition and faith in the "mirage realm" of consciousness neither denies nor belittles the importance of those experienced realities that cannot be science-quantified. Everything that is qualitative—including the aesthetic and ethical—falls in or near the faith category. But mirages, too, take their insubstantial forms from the common ground of Nature. To understand the material world as fully as possible via the socially shared, measured knowledge that is science, while maintaining an openness to the immeasurable, mysterious and wonder-filled experiences of existence on Earth, is an attainable goal. Granted, keeping a steady focus on greater-than-human Nature is difficult, which is something of a paradox for the Earthlings that obviously we are. Modern culture's narratives tend inward, toward human-centredness and self-centredness, inclined away from Earth. Wordsworth's lines put it simply and truthfully:

> Getting and spending we lay waste our powers
> Little we see in Nature that is ours.

This essay concerns ecological, aesthetic and ethical implications of Earth's life-forms evolving from stardust—of humans-as-Earthlings parallel in lineage with everything perceived in their surroundings back to the beginning of time. Human ancestry is a branch of animal ancestry, but with the added dimension of complex culture that shapes consciousness. Culture, transmitted by discourse, can be harmonious with reality or discordant. A critical question today is whether and by what means an affectionate, sustainable Culture of Earth can be developed from the questionable culture of homocentric individualism that currently reigns.

Civilization will not survive without an Earth ethic. How can the arts and sciences contribute? Perceiving our selves in the context of the planet, as one animated class of its vital parts, is a good start. An additional incentive is the promise that allegiance to Earth will transcend and moderate ethnic-cultural loyalties, the source of jingoism and other totalitarian forms of chauvinistic nationalism.

Aspects of Selfhood

For two thousand years and more humans have been exhorted by their spiritual leaders "to be better." The mayhem of the last century—one hundred to one hundred and seventy million killed in wars and strife—indicates that such preaching is fundamentally flawed. Is there no cure for the homicidal pathology that pads the pages of history? Perhaps the genetic temperament of *Homo sapiens* is an evolutionary error; the tumorous frontal lobes of the human brain may be analogous to the Irish elk's antlers, which, steadily evolving in size, dragged it down to extinction. Another theory, suggested by Arthur Koestler, is that the inherent *Homo* defect is failure of the brain to integrate successive stages as it evolved from reptile to mammal to human. Today, on the psychiatrist's couch, the analyst is presented with the perplexing problem of a crocodile, a horse, and a human vying for expression in the same personality. If true, the only hope for a peaceable world, he said, is the universal prescription of a soma-cocktail of pacifying drugs and hormones.[1]

Optimistically, the human flaw may not be inborn. It may be something correctable without gene therapy, such as misunderstandings of individual and group psychology, lack of attention to cultural myths, confusion as to the relative validity of rational versus intuitive promptings of the mind, ignorance of human origins and dependencies, or the imperfections of social institutions. If so, education is the hope, and where to begin the first question.

Ken Wilber, transpersonal psychologist/philosopher, has developed a fanciful scheme of human advancement around a reasonable idea: that changes in the human condition necessarily involve development of four closely interrelated aspects or "dimensions" integral to selfhood: two mental (interior and interpretive) and two physical (exterior and empirical).[2] The interior/mental dimensions are (1) the ego "I" or body/mind with its inherited biological intelligence, and (2) the collective ideas and values of the "we" culture: the shared worldview. Matched to the interior "I" is (3) the exterior non-humanized world—the natural or empirical "it" world that each ego perceives and interprets. Matched to the interior cultural "we" is (4) the material environment of planned landscapes, architecture, roads, etc.—the social "it" of institutions that outwardly expresses culture's political, economic, legal, medical, and educational agenda.

Wilber draws the useful distinction between cultural (ideational) constructs and social (built) constructs, the two constituting the fundamental framework of every civilization. When to this cultural/social foundation humanity's shared biological heritage and ecological connections to Earth are added, then the idea of the I-ego is placed in correct perspective. The body/mind differences that separate individuals from one another are small compared to the commonalities they share.

Each of the four aspects contributing to selfhood has its appropriate disciplinary study or studies. Psychology seeks understanding of the "I," while the natural sciences examine the empirical "it" world discerned by the ego. Certain social sciences (anthropology and sociology) concentrate on the cultural "we," while others (sociology again) study culture's outward manifestations: the built "it" environment. Although this four-part taxonomy is an attempt to understand a seamless web of relationships, each aspect has been seized on by champions crying, "I have found the Way! Follow me!"[3] Gurus such as Krishnamurti (and Wilber!) promise the new day dawning when sufficient I-egos have been reformed. Devotees of scientism such as E.O. Wilson insist that the "It" and "we" worlds must be perceived reductively in terms of the laws of physics. The cultural "we" attracts the attention of anthropologists, linguists, and educators such as C.A. Bowers. Politicians, influenced by Marx, are mostly intent on changing the "it" of social institutions and their supportive concepts.

Like the fabled blind men, each group has grasped one of the quadruped's supports and is pushing and pulling on it, at cross-purposes with the others. Some degree of co-ordination, so far lacking, is necessary for forward movement. The cultural "we" gets least attention, therefore its emphasis in this article.

Modernity spotlights the individual, encouraging a focus on the I-ego and its reform. The crowd of attendants on this first appendage is large, and advice on how to move it along the right track is prolix—as attested by the numerous books in the New Age sections of bookstore shelves—"Change personal consciousness!" This popular chorus ignores all other means of advance. How often do we hear: "In the final analysis, it's up to the individual to change!" Centred on the "I" part of selfhood, the message is harmonious with contemporary thought and diverts attention from serious cultural, social, and environmental issues. The thesis that only spiritual growth, self-improvement, self-realization, self-development, self-acceptance will change the world is doubly advantageous because it shames the weak-willed common herd while gratifying those who benefit from the status quo. The latter are happy to have the imperfections of the consumption culture blamed on individuals in general. They approve Pogo's, "We have met the enemy and s/he is us" that unjustly spreads the blame over all. People are encouraged to think themselves sinners rather than to damn the system that makes them so. Individualism is foregrounded, culture is backgrounded.

Individualism as a Cultural Construct

Humans in general are semi-gregarious animals, neither loners like tigers and moose nor obligate flock-members like ants and honeybees. "Semi-gregarious" is an on-average description, because the graph of any human population on the egregious/gregarious baseline is undoubtedly bell-shaped. At the left extreme are those who are unhappy except when in crowds; at the right extreme are recluses and hermits. The marvel is that the vast majority—the semi-gregarious under the hump of the bell—have been culture-convinced by neo-Darwinian thought that their nature is to be competitive, aggressive, "rugged individualists" on the far right, despite the mutual attraction that draws them together in families and communities, in town and city living. According to current ideology, everyone ought to be egregious, special, creative, autonomous, a stand-out from the crowd—which bespeaks the hidden power of culture to lead people hither or yon. The bell curve of human beliefs and behavior has been moved asymmetrically to the right, seriously weakening leftward trends that point toward Nature's communal realities.

Charles Taylor examined the history of true-to-oneself individuality, treating it as a commendable psychological development rather than as a questionable cultural construct.[4] Modern individuality begins with Descarte's idea of disengaged rationality—everyone should think for themselves. Hobbes and Locke established the political individual as prior to and more important than society. The Reformation taught that persons could find God's moral promptings within their own consciences. Then, as theistic beliefs faded, these promptings were attributed to human nature, viewed by Rousseau as the trustworthy source of ethical actions liable to corruption by society. The idea of self-determining freedom followed, interpreted by Herder as each individual having an original way of being human. "This above all, to thine own self be true." Personal authenticity was born with the thoroughly modern creed, "My calling is to live my life in my own unique way and not in imitation of anyone else's." Or as Sinatra sang it, expressing what every patriarch ought to say with satisfaction at life's end, "I did it my way!"

Here, clearly, is a four hundred-year-old cultural construct, a story of ourselves that shapes our selves. Taylor argues its positive nature. The ideal of personal authenticity has identified one of humanity's important potentials. It calls us to work out our own opinions and beliefs. It points toward a fuller, more self-responsible, more differentiated form of life appropriated as our own. It allows a richer mode of existence. Further, he notes that authenticity rightly understood is not self-indulgence, not narcissism, because the genesis of personal identity is dialogical, developed against a background of "things that matter." We define ourselves through acquired "languages of expression"—of word-sentences, of art, of gesture, of love—and are inducted into these languages by "significant others." Trumping dissent, he points out that the cultural mainstream of Western liberal society is centred on this and other forms of individualism. Therefore, do not try to eradicate the ideal but espouse it at its best. But George Grant identified the central weakness of "liberal society". "I mean by liberalism," he said, "a set of beliefs which proceed from the central assumption that man's essence is his freedom and therefore what chiefly concerns man in this life *is to shape the world as we want it*"[5](emphasis added).

The problem with personal authenticity, as with many other humanist ideals, is that it pays little attention to the importance of the Earth context. It is caught in the blind alley of homocentrism. The "things that matter" to it are virtually unchanged from what they were one hundred years ago. Taylor mentions a few new "significant others"—natural surroundings and wilderness—but little is made of them. According to him, the cultural mainstream of Western liberal society only needs fine-tuning. He believes that the egoistic imperfections of personal authenticity are not intrinsic, but rather are forced on it by the outside pressures of an atomizing society, by individualistic ideas of mobility, by entrepreneurship, by the self-centredness of "popular culture," by the nihilism of "high postmodern culture." Thus he sees "authenticity" of the individual as an end in itself, not as a means contributing in some way to sustainable living on Earth. When he writes vaguely that "the culture of self-fulfilment" requires unconditional relationships and moral demands beyond the self, this reader cannot but wearily think, "Another vaporous exhortation to be good or at least better!"

The individualistic culture, the culture of self-fulfilment and personal authenticity, is necessarily in-turned no matter what high ideals it espouses. Whatever altruism, ethics, and aesthetics it approves will be locked into a humans-only world that takes "man as the measure." Here some may argue that, as human beings, we cannot do other than cast our perceptions and value systems *homo/morphically*, in human forms, and that is correct. But to conclude from this fact that our perceptions and value systems must necessarily be centred on humanity, *homo/centrically*, is incorrect as many religions show. An aesthetic and ethical vision beyond individuals, beyond human society, is possible. Pointing out that "We are living under a cultural hypothesis," artist Suzi Gablik commented, "The problem with modernism, as I perceive it now, is that it has failed to understand individuality as a stage, a temporary subjective condition—something to be cultivated *for the sake of something else.*"[6]

The hinted at "something else" must transcend homocentric society and community because neither of these aggregates is a self-sufficient, self-standing entity. Both "community" and "society" are taxonomic categories, organic groups abstracted from context. They lack functional existence or living-meaning, unless perceived as embedded in one of Earth's supportive ecosystems. Gablik's "something else" of importance is that which envelops and encapsulates humanity. Various nature philosophers are calling attention to this reality, to the ecological dimension that has been marginalized and relegated to the merely "mundane" by the revealed, scriptural man/God religions. Ecological ethicists are attempting to surmount the "culture of individualism" and establish humanity's common ground in Earth. Their claim that the future of *Homo sapiens* depends on development of a new kind of ethics—Earth ethics or eco-ethics—seems to me irrefutable.

Philosophies of Nature

The philosopher Santayana asked the questions: Why do the world religions shy away from Nature? Why has Western man's conscience rebelled against naturalism and reverted in some form or other to the cult of the unseen?[7] Usual answers are the precariousness of life in agriculture-dependent societies, the inability of men to relate to their bodies and to Earth in feeling/emotive ways that come more easily to women, the general imperfections of existence, the facts of human anxiety, suffering and death, and therefore the hope for an unseen power to establish dependable order and permanency in the midst of flux. The old hymn summed it up, "Change and decay in all around I see, Oh God who changeth not, abide with me."

Theologians and mystics have toted up the score and concluded that the ills of life outweigh its joys, even when to the latter the ever-renewing beauties of Nature are added. Turning inward, they have explored deep consciousness and found evidence of an other-than-Earthly existence free of this world's woes. Philosophers too have traditionally treated aesthetics—the healthy engagement of feelings and emotions with Earth's web of relationships—as marginal to the man/God connection, as borderline to the important problems of reality and how it is known. Nature philosophers have been few—at least until modern times with the likes of Gregory Bateson and Arne Naess—which in part explains why Earth has been treated so shabbily.

Bateson thought that humanity's profound religious need was to affirm membership in what he called "the ecological tautology," meaning the eternal verities of life and environment here on Earth.[8] Ecopsychologists back him up, naming the prevalent human psychosis as EDD—Earth Deficiency Disease.[9] Like Santayana, Bateson asked, why do people associate the satisfaction of their "spiritual need" with extra-sensory perception, with spiritualism, with immaterialization and out-of-body experiences? These are mistaken attempts, he suggested, to find alternatives to an environmental crudeness that has become intolerable. They represent efforts to escape from a reality that has been degraded and uglified by industrial civilization. "A miracle," he said, "is a materialist's idea of how to escape from her/his materialism," adding that the true reply to crude materialism is beauty. Thus he drew the useful distinction between a "perceptive materialism" that accepts the mystery and beauty of the physical universe, and a "crude materialism" of careless industrial exploitation that in its haste to increase human wealth at Nature's expense destroys her harmonies and beauties.

If "crude materialism" is the enemy, and beauty the escape route, then "crude materialism" must be replaced by "beautiful materialism." Where but in Nature do we find it? Not the Nature of industrial civilizations and cities but the Nature that has so far escaped human domestication, simplification and uglification. This beauty-in-Nature is the rallying call of all current "environmental" (better "ecological") activities ranged against the excesses of the commercial market society. Ecological understanding has expanded the scope of movements that began with the intention of saving the whales, the polar bears, the prairie orchids, and thousands of other marvellous creatures, to include their habitats, their geographic ecosystems—the exquisite landscapes and waterscapes with which they evolved and to which they are adapted.

The primary motivation of the ecology movement issues from the innate aesthetic sense and the ethic that flows naturally from it. When the Environmental Advisory Council of Canada's federal government sought the basis for an "environmental ethic," the sturdiest hook they could find to hang it on was sensuous beauty-in-Nature.[10] Opposed to it is the proclivity to turn inward, to seek the aesthetic out of context, to redefine comeliness symbolically as "diversity" or as intellectual elegance—"the beauty of mathematics" (which is more fragrant, $E = MC^2$ or a rose?) Perhaps ecological understanding that emphasizes the importance of Nature's surround, combined with spontaneous aesthetic experiences there, can enlarge one's idea of "self" to include Earth's life-forms in their land and water ecosystems? This is Arne Naess's thesis.

The Ecological Self

Naess, the founder of the Deep Ecology Movement, is an eminent philosopher, a mountaineer and lover of the wild. His call for action has been motivated by what he sees as the appalling deterioration of planet Earth, overpopulated and under attack by a consumer society. The intrinsic importance of Nature as the locus of reality and, therefore, the source of ethics and aesthetics, is central for Naess. Concerning search for the supernatural he has remarked, "it can easily become an endeavour hostile to man and environment, though it can lead to splendid art!"

Naess is a realist, accepting the world as perceived. He remarks that a couple of thousand years of philosophy have not settled the identity of "my self," though traditionally people have talked of an "ego-self" that develops into a "social-self" and then a "metaphysical-self," with no necessary relationship to Nature. But from the beginning, everyone is an "ecological self"—in, of, and for Nature. Let us therefore adopt the appropriate name, *Homo ecologicus*.[11]

How large is one's "ecological self?" Naess answers with another question: With what can you *identify*? The paradigm situation of *identification* is compassion, love, feeling for the other. Thus the "ecological self" is far richer than human society; it can encompass with empathic feelings the wider web of relationships that exist between humans and the Ecosphere. It can be expanded throughout a lifetime. Conducive to the widening and deepening of ecological awareness is "dwelling in situations of intrinsic value, relaxed from striving, open to perceptions of spontaneous, non-directed (Gestalt) experiences." This is doubtless an apt description of Naess at his Tvergastein mountain-top retreat in Norway. To be "out in Nature" is for him to be "in his ecological-self."

The richness of personal spontaneous experiences in Nature are not subjective, says Naess. The qualities sensed in a grove of golden-leafed birches rustling and swaying in the autumn wind are in the identity of the trees, and beauty is *not* just in the eye of the beholder. The secondary and tertiary qualities registered by our senses—fragrances, harmonies, sympathies—are realities in the world, more real than the so-called primary attributes of mass, weight, volume, etc. that are the measured abstractions of science.[12] Ecological-self realization, embracing society and all Earth's life-forms, opens the human to the joys and love of Nature—from which flow altruism, unselfishness, and what Kant called "beautiful acts." No moral exhortations are needed to prompt care for what the ecological-self embraces. In short, Naess argues that worthy environmental ethics are derivative from the first priority, which is to gain a worldview that expands the ego-self to the ecological self at one with Nature.

Naess's idea that a comprehension of Earth reality (ontology), its delights and beauties (aesthetics), will engender right environmental actions (ethics), strikes me as authentic. But in a critical vein, if one begins with the ego-self and gradually extends its boundaries to become the ecological self, widening the circle of concern to include more and more of humanity and Nature, then one necessarily subscribes to "ethics by extension." Just as illumination decreases as the square of the distance from its source, care and affection obey the same law of diminishing intensity. The vigor and strength of ethical concern weakens rapidly as it extends outward from self to family to friends to animals and plants, to landscape ecosystems, to encompassing Nature. A better approach, it seems to me, is to identify Earth initially as the central reality and source of value, from there working inward to establish the value of its parts as contributors to the whole. This "ethics by inclusion," an eco-ethics that sets humanity in the perspective of the planetary worldview, is the more promising track. A second criticism of Naess is that the importance of cultural myths as impediments to ecological awareness seems to have escaped his close attention. His is an individualistic philosophy.

Ecologizing Culture

In its broadest sense, "culture" is social learning, the knowledge and habits acquired from others. In terms of conscious awareness, culture has various depths between the obvious and the hidden. The "shallows" comprise the visible characteristics of society, the ways of knowing and of doing, the arts and sciences. The "deeps" are the axiomatic, taken-for-granted, invisible beliefs, attitudes, and activities passed on from generation to generation by all the grammars of language and body-language, giving the social group (usually the nation) its particular idiom and cohesiveness. All levels of culture are connected; the visible aspects are expressions of the invisible. For example, the current fad of female skinniness is a comment on the ideology of high consumption whose other face—the ideology of the market—is reflected in the commodity forms taken by contemporary arts and sciences. As previously noted, the ideology of "individual authenticity" though historically recent is now an embedded part of Western culture.

At its deepest level, culture is an interlocking system of myths carried and transmitted by discourse, i.e. by all the signs, symbols, vocal and behavioural languages through which members of the same culture communicate. The myths are socially cohesive, giving people a common sense of history, of who they are, their purpose in being in the world, and where (optimistically) they are going. Language is the chief carrier of culture, enriching it with the cumulative knowledge and sensibilities of the past as well as with imaginings of the future. Girls and boys raised by wolves without language would be "wolf-children" barking and running around on hands and knees, or at best the equivalent of our most intelligent primate cousins, the bonobos and chimpanzees. But complex language-cultures, engrafted on the biological heritage, by "speaking us" transport humanity to a different communal plane from most other animals. Thus each separate body-self carries a shared consciousness, and the individual personality is like one of a multitude of visible grass stems that spring from a common underground rhizome system. The particular talent of each body/mind is a relatively small biographical "spin" on the surface of the biological and cultural deeps that bind the discourse-group together.

Bowers, champion of the "cultural self," challenges Naess's proposition that individuals can go straight to the heart of reality through spontaneous experience, bypassing culture, working out their ecosophies, enlarging their "ecological selves," and thereby promoting environmentally sensitive social change.[13] He approves Naess's assertion of people's ecological ties to Earth, but asks: How can the goal be reached without first dragging into the light and correcting the deranged cultural myths that lurk in the depths of the popular mind, framing and dictating most of their convictions? Why, for example, does the vaunted "reason" of the individual take a holiday when, as on September 11th, the national myth is challenged?

Bowers is an educator. His major theme is that teachers must understand the power of language/culture's entrenched myths to block changes that could improve the human-Earth relationship.[14] Consider how language with its baggage of outworn concepts moulds thoughts and actions! Consider how the media, the tools of communication—especially TV and computers—isolate and strengthen a false individuality! Consider how conceiving creativity as issuing primarily from the depths of individuals rather than from their ecological relationships denigrates society's culture and Earth! Before progress can be made toward sustainable living on Earth, the most dangerous cultural myths must be replaced. The in-turned presuppositions of a homocentric, individualistic society must be challenged and turned outward. Culture must be ecologized.

Obsolete cultural assumptions are many. Here I comment on two closely related ones:

1. Myth of Homocentrism, that humanity is central in the universe, the best of life, the sole locus of values, and that all Earth is a resource for human use.
2. Myth of the Autonomous Person, elevating the intelligence, creativity, moral judgment and experience of individuals over the same attributes of culture, with total neglect of Earth intelligence and creativity.

Earth is the Centre of Value—Myth of Homocentrism Denied

Each human being is a biological/cultural person whose selfhood and consciousness is accreted, life-long, through a web of relationships. Undeniably we are ecological beings, *Homo ecologicus*, and education should realistically orient everyone in that direction. A new name is the place to start. With the same facility that science-fiction writers identify beings from Mars as Martians, so people should realistically identify themselves as *Earthlings*—evolved out of Earth, born out of Earth, maintained daily by Earth, with total dependence on Earth for their lives.

No known living things exist apart from Earth. Any organism thrown into outer-space without a surround of Earth-habitat—atmospheric gases, water, food nutrients, air-pressure—is soon dead. The vitality of organisms is maintained by Earth from which they evolved. The idea that Organism = Life is the "biological fallacy," based on erroneous thinking-out-of-context. By focusing attention on organic "things like us," it helps to maintain the fiction that humans are the central fact of creation. Realistically, the organic and inorganic parts of Earth are inseparable, and from that animated matrix everything existent, including humanity, has been derived. Therefore, Earth is primary, organisms secondary, humans tertiary.

The metaphorical nature of language fosters misconceptions, such as the biological fallacy. The dictionary meanings and derivations of abstract words often show them rooted metaphorically in the way humans exist physically on Earth. Innumerable terms incorporate the ideas of in and out, over and under, up and down, back and front, together and apart. For example, mental phenomena expressed as com/prehension (grasping the whole), super/stition (standing above), under/standing (standing below), clearly show their body-in-Nature roots. Thus physical realities have provided the metaphorical means of describing immaterial concepts. Just as "heart" is the metaphor for "courage," and "brain" the metaphor for "mind," so in pre-ecological society "organism" was adopted as the physical metaphor for "life."

In the new world of ecological understanding, a more appropriate metaphor is Earth = Life. The miracle of life is in the Earth home, not isolated in its parts. *Media morte in vita sumus*—in the midst of death we are surrounded by life! The implications are many: homocentrism is a false myth. Humanity is not central in the universe, neither the best of life nor the sole locus of values. Earth has its own intrinsic values; it is not a resource for human use. Therefore, a primary human goal is to seek ways of living sustainably on Earth, contributing to its healthy diversity and beauty, not as the commanding executive officer but as the junior partner.

The Ecological Individual—Myth of the Autonomous Person Denied

In all the arts and sciences, in politics, economics and education, in the historical records of the various accomplishments of the species, the "creative individual" wins the victor's garland and earns humanity's

Nobel Prize. These select winners—lucky in heredity and environmental upbringing—are gifted with the ability to weave the threads of cultural discourse, thought, and imagination in novel ways. When the results are perceived as aesthetically appealing and socially beneficial, fame is their reward. Henceforth they become role models, patterns of success for the young, the paradigm of the "creative individual" to which everyone aspires. The importance of the autonomous, authentic individual is re-emphasized, the ecological individual eclipsed.

The ties joining the adjective "creative" to the noun "individual" are tenuous. Granted, the ego/I is a contributor—though a relatively small one—to whatever originality the individual may display over a lifetime. But intuitive creativeness, the "aha" experience, is not an item of personal ownership and, recent legal precedents to the contrary, the discoveries and inventions of individuals are not "intellectual property." The largest part of "selfhood" is drawn from the biological/cultural commons. Everyone inherits a human nature that confers the evolutionary body/mind wisdom of the race: a collective wisdom adapted to surviving on Earth. The deeps of unconscious knowledge, the emotions and archetypal images, are also common property. To these are added the aggregate influences of discourse, of signs, symbols and language, as well as taught perceptions of Earth-reality—all within the formative context of social institutions. The idea of a stand-alone, creative individual is false. Yet a primary aim of the educational system, from K to U, is to implant firmly this nonsensical concept in every student's head. The artist is the model.

The more people imagine themselves autonomous, the more alien Nature seems to be. Art sets the pattern, following culture in the sense that the artist is inescapably saddled with the same cultural baggage as everyone else. The innovative writer, poet, painter, sculptor, musician, is modernity's primary exemplar of the supposed authentic individual who draws inspiration and creativity from inner depths. Psychology, literature, science, education—all take the lone artist as the standard. Children are encouraged to express their own individuality, to generate novel ideas, to learn and evaluate independently, to find their own artistic souls and follow their bliss.

The X generation has been thoroughly infected with the idea of individual specialness. As a painful auditory example, consider popular music. It used to be that those with talents for composing did just that in New York's Tin-Pan Alley, and those with talents for performing were satisfied to perform. Not any more. Today's popular performers rarely do the melodious "standards" written by talented professionals. They must themselves compose, demonstrate their creativity, play or sing their own compositions. The result, for the most part, is execrable music and words to match. When the goal and primary mark of the artist is originality—that is, being different from others, absent any aesthetic and ethical context—the results are bound to be bad or at least indifferent.

In his book *Grammars of Creation*, George Steiner examined in detail the springs of creativity (poiesis, originality, discovery, invention, innovation) of those engaged in the arts, with peripheral attention to the sciences.[15] Immediately one asks, why so much scholarship devoted to a subject that on the surface appears trivial? The answer, it seems to me, is that *the prototype Artist of the modern era is nothing—unless creative in an individualistic way.* Hence the worried search within for metaphysical sources of inspiration, a search that proves disappointing. Steiner admits in passing that poiesis is crucially collaborative, that "all art is combinatorial and the concept of originality little more than a self-flattering illusion," and that "the most solitary of creative acts have a shared collective fabric." Nevertheless his heart is with *auctoritas* and not *communitas,* with the autistic rather than with the cultural-communal, and he writes, "in the arts, music, philosophy, and literature, solitude and singularity are of the essence. The creative motion is as individual, entrenched in the citadel of self, as is one's own death. Kinship of poiesis and death is their aloneness. We create and we die in ontological isolation." The spotlight is on the artist-as-lonesome-ego, exemplary both in living and in dying.

The Ecological Artist

Creativity is more than the originality that enhances individuality. The difference is in the vision of what is important and the source of that importance. The creative ecological artist "makes special" those activities, metaphors, rituals that strengthen relationships between humans and the natural systems on which they depend,[16] looking outward to the creativity of Earth from which, miraculously, all things perceived today have evolved. Creativity is in Earth's parts too—in water, air, soils, sediments, and in organisms. Humans are enveloped by a creativity in which they participate through their Earth-made biological natures and through their Earth-derived discourse-cultures. The originalities of even the most talented individuals are small variations in the knots of Nature's vast ecological net. The aesthetic and ethical implications of the ecological perspective point new directions for both the sciences and the arts.

Current criticism of the old eases acceptance of the new. Steiner's pessimistic perspective on arts and artists can be read in part as an indictment of individualism, although this is certainly not his main thesis. In contrasting the individuality of artists with the communitarianism of scientists, he finds in the latter more hope for the future. The Australian philosopher Gare takes artists and writers to task for detaching people from their cultural roots, for attacking all forms of internal and external constraints, for advocating an aesthetic justification for life—to selfishly create themselves as works of art. In this they have been successful. Modernist artists and writers are no longer significant in society, he says, because they have won, there is nothing left for them to do.[17] Artists will disagree, justifying their work (as do scientists) by promising a wide spectrum of social payoffs. An example is one proposed by Hickey—the disorientation of difficult modern art, he argues, helps people to adapt to the anxieties of urban living.[18]

Given the freedom to choose, people spend their lives doing what most interests them. Rationalizations follow, and cosmic significance or at least cultural leadership is not uncommonly attributed to the chosen paths. As the literature of art criticism and theory affirms, artists are no exception to these universal rules. Further, some of those who speak for artists expect them to be leaders. Yet, *pace* McLuhan, those who love to paint and sculpt and make music and write are not born to be critics of their cultures, nor antennae of civilization's future. Artists are known as such because they express the human aesthetic sense in ways that resonate with the senses, feelings, and emotions of others. Like everyone else, artists are particles in the flow of the Nature/culture stream; they are no more to be looked on as prophets than to be blamed for the imperfections of the philosophy/culture that carries them along.

What merits criticism is the underlying philosophy of the arts, and of the sciences, that pins humanity's hopes on the deep potentialities of the authentic individual. The times suggest that these inner explorations have reached their dangerous limits. The next avenue of discovery opens outward, to an aesthetic perception of Nature as the creative source. Here, also, an ethical altruism will be discovered that includes while it transcends the human race.

References

[1] Koestler, Arthur, *Janus, A Summing Up* (New York: Random House, 1978) 9, 20.

[2] Wilber, Ken *A Brief History of Everything*, 2nd edition (Boston: Shambhala, 2000) 63-75.

[3] Ibid., 77.

[4] Taylor, Charles, *The Malaise of Modernity* (Toronto: Stoddart, 1991).

[5] Grant, George, "The University Curriculum," *The University Game*, ed. Howard Adelman and Dennis Lee (Toronto: Anansi, 1968) 48, footnote (3).

[6] Gablik, Suzi, "Changing Paradigms for the Artist," *The Structurist* 25/26 (Saskatoon: University of Saskatchewan, 1985/86) 25-29.

[7] Durant, Will, *The Story of Philosophy* (Toronto: Doubleday, Doran & Gundy, Ltd., 1926) 537-538.

[8] Bateson, Gregory, *Mind and Nature, a Necessary Unity* (Toronto and New York: Bantam Books, 1980) 232.

[9] Morse, Norman H., *An Environmental Ethic—Its Formulation and Implications* (Ottawa, Department of Environment: Canadian Environmental Advisory Council, Report No. 2, 1975).

[10] Ibid., Morse.

[11] Naess, Arne, "Self-realization: An Ecological Approach to Being in the World," *The Deep Ecology Movement, An Introductory Anthology,* eds. Alan Drengson and Yuichi Inoue (Berkeley: North Atlantic Books, 1995) 3-30.

[12] Naess, Arne, transl. and revised by David Rothenberg *Ecology, Community and Lifestyle* (New York: Cambridge University Press, 1989) 52-55.

[13] Bowers, C.A., *Educating for an Ecologically Sustainable Culture: Rethinking Moral Education, Creativity, Intelligence, and Other Modern Orthodoxies.* (Albany: SUNY Press. 1995) 169.

[14] Bowers, C.A., *The Culture of Denial: Why the Environmental Movement Needs a Strategy for Reforming Universities and Public Schools* (Albany: SUNY Press, 1997).

[15] Steiner, George, *Grammars of Creation* (New Haven and London: Yale University Press, 2001).

[16] Bowers, C.A., *Educating for an Ecologically Sustainable Culture: Rethinking Moral Education, Creativity, Intelligence, and Other Modern Orthodoxies.* (Albany: SUNY Press. 1995) 69.

[17] Arran E. Gare, *Postmodernism and the Environmental Crisis* (London and New York: Routledge. 1995). p. 16.

[18] Hickey, Dave, "A World Like Santa Barbara," *Harpers Magazine* September 2000, 11-16.

PROGRESS IS GREATER CONNECTEDNESS

Reprinted with permission from *The Structurist*, No. 37/38, 1997–1998. University of Saskatchewan, Saskatoon, pp 52–55. Originally titled "Progress and Connectedness." Edited for this collection.

The original meaning of "progress" is "walking onward," leaving the old behind in search of the better new. Assuming that "the better new" should be realistic and positive for life on Earth, then true progress is any advancement in ecological connectedness between ourselves, our artifacts, and the evolutionary Nature-matrix wherein we exist. At this nexus the contributions of our arts and sciences can be judged. In the words of Eli Bornstein, "Ultimately the cultural and ecological value of the education, history, and practice of art and architecture depends upon their connections with history and nature."

Here I examine some of the ways that we have separated ourselves from Earth's realities, mostly by privileging mind over body, ideas and concepts over direct perceptions, reason over intuition and feeling. Philosophies and religions, the arts and sciences, have often fractured rather than strengthened connections to the irreplaceable medium that sustains us all. Rather than cultivating symbiosis, they have fostered separation and made a virtue of it.

Separation: False Progress

It has been said that the only fundamental human sin is separation. Yet in the history of our race much of what is called progress has been a process of separation from the Nature that made us the particular primate species we are. Eight or ten millennia ago the adoption of sedentary agriculture with its simple coterie of tamed plants and animals ended intimate living with the diverse wild world that for several million years and more had been our ecological and evolutionary milieu. Again, urban living—accepted as progressive—is severed from the agriculture and aquaculture that make it possible; the cityscape is separated from the productive landscape. Within dense settlements technological invention has enclosed people in various kinds of insulating cocoons—buildings and vehicles—so that an estimated 95 per cent of the average adult's life is spent indoors. Virtual reality is the ultimate isolating technology: each individual bunkered in with a computer screen, wired to an ersatz universe, segregated from society and the world beyond it. Is each new stage of artifice an advance, or false progress?

A parallel severing of connections can also be traced in the history of ideas as to who we humans are. Twenty-five-hundred years ago our species—especially as epitomized by the male sex—was conceived as exceptional, ensouled, and therefore completely different from other animals. Not long thereafter it was disclosed to the revealed religions (Judaism, Christianity and Islam) that pantheism and animism, with their beliefs that life is immanent in all things of Earth, were false creeds. For if the world is alive, and everything in it an aspect of one global Life, then human society is simply a minor part with no special rights. How much better to believe in a transcendent God whose special concern is the human race, to believe that humans alone are worthy of immortality, their true home, heaven and not this Earthly veil of tears.

Denial of ecological reality is a chasm that Darwin's theory of evolution has not bridged. The fundamentalists who disavow the evidence of evolution are only part of a deeper problem that stems from the ancient myth of humanity's extraordinary uniqueness and separateness. The theme of special worth, devaluing by comparison everything else in the world, is so deeply and fundamentally enshrined in the secular ideas of modernity that humanity's Earth beginnings and animal ancestry can be accepted without diminishing one whit today's exploitive attacks on Earth and its non-human constituents. Are we not the highest point of evolution, a distillation of mind and spirit far above the animal world, and by that fact entitled to take whatever we want? Perceiving the world as apart from us provides the legitimacy for this comforting faith.

Perceptions of reality are guided by philosophic beliefs, and Matson divided them into two basic streams: the *inside-out* exemplified by Descartes' approach that begins with the data of human consciousness and goes on to explain the world, and the *outside-in* exemplified by Spinoza's approach that begins with an account of the world and then narrows in to an explanation of the mind and its knowledge.[1] The first is attractive to theological reason, the second to ecological reason. But choice of philosophy should not be just a matter of aesthetic appeal influenced by temperament and training, no more than a personal viewpoint that, as such, cannot be challenged. The belief systems by which we order our lives and the lives of others are critically important. Is there a way to sort out different ideas, different worldviews, in order to judge their worthiness? Attention to the context within which they are appraised can be helpful.

Context Clarifies Conundrums

In an essay titled "Two Types of Problems," Schumacher distinguished between what he called "convergent problems," such as the best design for a bicycle, and "divergent problems," such as how to maintain an ordered society that nonetheless supports freedom.[2] Solvable mechanical problems typify the first set and, for example, ideas have *converged* worldwide on the one "best bicycle" design. The second set, typified by difficult social problems, elicits *diverging* ideas, each attractive in its own way, that suspend easy solutions.

Consider the question of how best to educate children. A good case can be made both for an authoritarian approach that requires discipline and obedience on the part of the students, and for a facilitation approach that allows the students the greatest possible freedom. Which should rule for the benefit of the children, discipline or freedom, when both are desirable to some degree? The solution, wrote Schumacher, must be one that both engages and transcends the two opposites—great teachers "must *love* the little horrors." An atmosphere of love, sympathy, and compassion in the school fosters the wisdom to weigh the degree of freedom and the degree of obedience appropriate for each educational situation. The solution takes its cue from a "higher" vantage point, one that transcends yet merges the diverging concepts.

The idea of finding solutions to problems by going outside their boundaries, by transcending them, is equivalent to widening the context, broadening the frame of reference in which they are assessed. Abortion is a typical "divergent" problem, the cause of much social discord because it poses choices between potential child and potential mother. In the strict humanist context this problem is apparently unsolvable. But when the question is posed in the transcendent context of Earth, already overcrowded with its 6.5 billion humans going for nine billion by 2050, abortion (when conception control fails) seems an eminently reasonable choice, worthy of the broadest social endorsement and support. A healthy Earth is more important than a sick, overpopulated one.

Whether to guide our lives and our actions according to some form of realism/objectivism or idealism/subjectivism is a "divergent" problem in philosophy and theology. As with the practical problem of abortion, this question of theory is ambiguous at the social level though less so, I believe, when set in the Earth context. Judgments as to the values of different philosophies and theologies require attention not only to their likely outcomes in society but also—more importantly—to their practical outcome on Earth. Do they make connections? Are they progressive?

In Praise of Realism

John Ralston Saul is a champion of tough social realism. Philosophy, to him, should be a political tool helpful in rediscovering ourselves and in shedding the linguistic obscurantism of whatever power is in place. His aversions include the current ideology of economic determinism, preaching the faith that "the bottom line" necessarily rules the world, and scholasticism as academic bean-counting instead of critical and vociferous doubting a la Voltaire. Deconstructionism and uncurbed Reason are two of the ideas he labels dangerous and concerning which—as befits a genuine doubter—he has pertinent things to say.[3]

Postmodern deconstruction or deconstructionism is a philosophy of linguistic idealism. It asserts that language makes reality or rather there is no "real," only social constructions of language and concepts that frame and structure all human experience. Saul interprets this notion as arguing that human communications have no inherent ethical, creative or social worth. Thus the value of public discourse is undermined, playing into the hands of whatever anti-democratic forces are at work. As "a generalized denial of civilization," writes Saul, "deconstructionism can't help but be a voice of evil. . . . It is a prime example of how intellectuals [mostly academic] create anti-intellectualism."[4]

If "linguistic idealism" is disputable at the social level, how does it fare in the broader context of Earth? Here Saul has little to say. Charlene Spretnak, an "ecological postmodernist," sounds the alarm against both modernity and its supposed antithesis, postmodern deconstructionism.[5] The modernity she castigates is roughly the equivalent of Saul's "corporatist society," a threat to the encompassing ecological values of the natural world for which it substitutes the abstract, the symbolic, the synthetic. Deconstructionism carries the abstract even farther. By insisting that language makes reality, that "Nature is what we choose to make it," this form of idealism imposes devastating discontinuities between people and the places where they live in the wider world. The curse of deconstructionism is to assert that there never were real connections to be perceived but only ephemeral ties, these latter the inventions of culture in the interests of obscure power plays. Judged on the basis of "making connections," this brand of idealism strikes out.

To be true to our biological/ecological history we must embrace some form of Earth-based realism, the more so when we find ourselves breathing acrid air, eating vegetables preserved with methyl bromide, bathing in chlorinated water, and for safety reasons drinking only bottled fluids. Reason as presented in the "pure form" of science claims to disclose the reality of the universe, to bring us closer to it rather than separating us from it. Certainly science has aided our understanding of human biology, evolution, and ecology, though not yet in a revolutionary way. Is more science the path of progress? Will objective reason draw the necessary connections?

The Trouble with Reason

Reason has a tendency to drift away from the immediately experienced world into realms of abstraction, but this is not Saul's complaint. He finds no fault with reason *per se* but with its pretensions of being the leading talent of mind, its subjugation of other equally important faculties, and society's obsessive treatment of it as an absolute value. Reason has become an ideology, a supreme truth that denies truth to other facets of mind, a justification for such "scientific" systems as Marxism, capitalism and corporatism—all staking their claim to rational truth. To those who protest that he is tilting against "instrumental reason" for which pure reason should not be condemned, Saul retorts that the attempt to divide reason into bad and good parts is subterfuge; there is only one "reason." Although science is not specifically named, it is undoubtedly the paradigm he has in mind when he writes:

> Reason might make more sense if it were relieved of its monotheistic aura and reintegrated into the broader humanist concept from which it escaped in search of greater glory in the 16th Century. In this larger view it would be balanced and restrained and given direction by other useful and perhaps also essential human characteristics such as common sense, intuition, memory, creativity and ethics. In such a generous context it would be easy to see that reason on its own is little more than a mechanism devoid of meaning, purpose or direction . . . in short, reason detached from the balancing qualities of Humanism is irrational."[6]

Amending the last sentence to "reason detached from the balancing qualities of an ecological Humanism is irrational," let us examine the question as to whether science "on its own is little more than a mechanism devoid of meaning, purpose or direction." If so, it must be a poor guide to progress.

Science, the Paradigm of Reason

When charges such as those of Saul are leveled against reason, especially when the word "science" is named, alarm bells ring in academia. A book by Gerald Holton illustrates the acute discomfort of the physical scientist when critics such as Vaclav Havel suggest that the shortcomings of "rationality" and "objectivity" have in this century contributed to the totalitarian terrors of the western world.[7]

Holton defends science in several ways:

1. By recounting all the good things that science has done for humanity,

including "the civilizing power of scientific thought."

2. By equating contemporary critics with past "romantics" (William Blake is specifically mentioned) who rejected "mechanistic materialism, rationalism, theory and abstraction, objectivity and specialization."
3. By denying that science has had anything to do with totalitarian dogmas such as Fascism and Marxism-Leninism (a book that Lenin wrote shows his ideas of science were faulty; also, Soviet scientists were mistreated in their own country—what further proof is needed?).
4. By arguing that scientists are not unaware of "the needed complementarity of mankind's rational, passionate, intuitive and spiritual functions," as exemplified in Einstein's scientific career and his philosophic writings.

The last argument would be telling were there more Einsteins in this world. As to his other points, Holton does not seem to understand the arguments of science's critics. He does not respond to the idea that reason tends to be hegemonic, claiming sole intellectual authority, dominating other important faculties of intellect and assigning lower status to them. Nor does he admit that scientists routinely adopt the political ideologies that environ them, contributing their skills as often to the powers of tyranny as to those of more benign regimes. He does not consider that science's way-of-knowing, its epistemology—for example, its espousal of "efficiency" and "parsimony" in the selection of hypotheses, as well as its mechanical metaphors and mathematical language—might contribute to confirming as normative such social forces as standardization, centralization, and regimentation. He does not recognize, as Einstein did in his later life, that science is instrumental, not an end but a means, not a "holy grail" of wisdom but a tool lending itself to control and management, amenable to good or evil purposes depending on the ethics of its users.

Scientists are not noted for a philosophic turn of mind. Rather they are apt to harbour a prejudice against speculative philosophy. In the words of Ziman, "One can be zealous for Science, and a splendidly successful research worker, without pretending to a clear and certain notion of what Science really is. In practice it does not seem to matter. . . . A rough and ready conventional wisdom will see him through." Ziman explains that this is because the scientist is involved in practical problem-solving using simple reasoning in ways "not very different from what we should use in an everyday careful discussion of an everyday problem."[8] The goal in science is to discover socially sharable knowledge, "consensable" knowledge, and not to ask, is science's traditional, realist/objectivist view of the world the one correct view? Has its favourite methodology of reductionism, anatomizing and atomizing things to get at control of their functions, caused any harm? Has it been progressive in the sense of mending separations, fostering ecological connections between viewer and viewed?"

The Objectivist View of Reality

The heavy hand of past activities and beliefs continues to shape the present. Air travellers flying over the western cordillera on clear days must gaze in dismay at the mayhem committed on the mountain forests by a civilized people. When custodians are asked about this legacy of skinned patches and scarred slopes the standard response is, "that's history, the old forest code; now we have the new forest code and practices are changed." Changed they may be, but the landscapes show no signs of it. Similarly with science, whose more thoughtful exponents such as Fritjof Capra protest that the old science code—separating viewer and viewed, subject and object, mind and matter—no longer rules.[9] Yet the legacy persists. Like the ravages of least-cost forestry, popular science as "objectivism" continues to blight social thought in the important fields of business, government, law, education and economics.

Most scientists are objectivists, believing that mind accurately mirrors the real world. Whatever they perceive and study is exactly what it seems to be. This faith, usually traced to Descartes and his separation of the seeing subject from the objective world, is an artifact of sight and touch. The visual and tactile senses pick out objects by contrast with their surroundings, differentiating figure from ground. Thus naive realism, with which science begins, postulates a reality "out there," totally independent of the seeing self. According to objectivism, the phenomena of the universe are to be understood by matching to them concepts and categories that in some miraculous way are the free creations of mind. True knowledge must not in any way be an artifact of thinking beings, but must represent what is really "out there" in the world.

When the experiences of all human minds are compared, the categories of thought that can be shared concerning the presumed objective universe are the rational, the logical, the quantitative. Everyone not blind can agree that a metre stick is one hundred centimetres in length, and that the distance from Earth to Moon is x thousand kilometres. The next "logical" step is to agree that the universe must be transcendently rational, logical, and quantitative in structure. If so all intimations of qualities, all intuitions and feelings, must be less authentic. They are not quantitative, not measurable. They are subjective, emotional, romantic, unconnected to the real world.

The Negative Aspects of Objectivism

Linguists Lakoff and Johnson outline the harmful influences of realism/objectivism.[10] Because it continues to be the philosophy of science, and because science remains the model for cognition, the latter is assumed to be purely rational, unemotional, detached, independent of imagination, of social functioning, and of the limitations of our bodies and memories. Then the quantitative and mathematical are glorified. More and more the computational and the mechanical enter everyone's life. Machines are given human attributes; some are said to have the ability to reason, to exhibit "artificial intelligence." Progress becomes the restructuring of people and their physical and social environments according to the efficient, mechanical, "scientific" mode.

The Industrial Revolution in Europe was the first socially disruptive result of objectivist reason, spawning its antithesis: extreme subjectivism. Reacting against the ugliness of factories and machinery, the rapid urbanization and destruction of rural living, such versatile artists as Coleridge, Wordsworth and Shelley, Goethe and Schiller voiced opposition to the science of the day and its fruits. They called for simpler living, exaltation of emotions and feelings, a focus on the intimate and spiritual, more attention to sensuous experience and poetic intuition, a return to nature. What is important, they said, is imagination not reason, inspiration not logic, personal experience not the impersonal and universal.

If this suite of impulses sounds "new agey," well so it is. The same reaction—this time against the second Industrial Revolution of scientific computerization, robotization and jobless growth—is with us today. When Vaclav Havel writes that man needs "individual spirituality, firsthand personal insight into things . . . and above all trust in his own subjectivity as his principal link with the subjectivity of the world," he hands ammunition to the science/objectivists who interpret such statements as a recrudescence of the irrational Romantic rebellion now linked dangerously to political theory and invading high office.[11]

The battle lines are drawn again, just as they were two centuries ago. At that time the Romantics opposed reason, championing imagination, emotion, and feelings as the road to higher truths. The result was exactly the opposite of what they had hoped.[12] Objectivism as empirical science marched on, recruiting to its side business, government, economics, law, education and the media. So far as political power is concerned the arts were marginalized along with religion, set aside as solaces at the end of the workday, scorned by the hard-headed as frills and curlicues, deemed dispensable when economies falter—unless they can be co-opted into the system and shown to have a money value. Clearly, a rapprochement between objectivist reason and subjectivist imagination is needed.

The Middle Way: Experientialism

Whitehead, the mathematician-philosopher, thought that an organic view of nature—by which he meant perceiving all parts as related—could heal the split, because then both qualitative and quantitative relationships would present themselves as "matter of fact" to the objectivist view. If reality is "organic," then aesthetic attainments and values must be accepted as integral to all events. Context would be perceived as fundamentally important for all organic systems. Whitehead did not view the Romantics as irrational. Their nature poetry, he said, is "a protest on behalf of the organic view of nature, and also a protest against the exclusion of value from the essence of matter of fact."[12]

In three books—*Metaphors We Live By*, *The Body in the Mind*, and *Women, Fire and Dangerous Things*—Lakoff and Johnson propose a different but complementary approach. They identify objectivism and subjectivism as ideologies confronting one another, thesis and antithesis, each needing the other's friction to maintain its own temperature. Their moderate philosophy—experientialism—borrows from both divergent extremes to effect a higher synthesis.[10,13,14]

Experientialism takes as its starting point the linguistic evidence that all conscious thought and feeling—scientific/rational, mathematical, imaginative, intuitive, aesthetic, emotional—is fundamentally metaphorical and that the metaphors, when traced to their origin, are derived from the relationships of our body/minds to the contextual Earth-reality in which we are immersed. In short, language is embodied and ecological. Rather than language making reality as some postmodernists argue, the reality of the body/mind/Nature relationship makes language.

Some of the simple "metaphors we live by," obviously related to our physicality, are up-and-down, inside-and-outside, ahead-and-behind, full-and-empty, more-and-less. Happy is up, sad is down. Good is up, good is more, good is ahead, good is full. Notice the linguistic difficulty in arguing that less rather than more might be progressive! Even emotional concepts are clear examples of abstractions having a basis in bodily experience. Anger is a rich category structure, its expressions paralleling the physiological effects of increased body heat and internal pressure, agitation, interference with accurate perception: "hot under the collar", "almost had a hemorrhage", "scarlet with rage", "so mad I couldn't see straight."

Experientialism agrees with subjectivism that intuition and imagination, the sources of creativity, are essential to the metaphorical conceptual structures by which, in every way, we understand ourselves and the world. Thus imagination and reason are alike based on the ecological relationships between body/mind and a real external world. Abstract thought—even mathematical thought—is derivative from the realities of common, shared human experiences. Understanding is by metaphorical projection from the concrete to the abstract. As with the arts so in the sciences imagination is central for finding connections, drawing inferences, solving problems. Subjective experiences establish the very structure of objectivity.

Experientialism also agrees with objectivism that a real objective world exists. But it insists that the way we perceive that reality is a human view not the God's-eye view. The dragonfly with its particular body physiology and its compound eyes of light sensitive tubes perceives a different world than we with our human single-lens eyes. Both worlds, the dragonfly's and the human's, are valid interpretations of "objective" reality. To grant that no aspect of our understanding is totally independent of the nature of the human organism, that reason is human and interpretive rather than objectively real and transcendental, that even mathematics is based on structures within the human conception system growing out of common bodily experiences, is not to surrender to irrationality. Within our world, as all humans perceive it, we can still arrive at the consensual knowledge called "science," not in an anything goes, purely relativistic sense but constrained by the greater surrounding reality that we can barely fathom. The Einsteins of this world need no longer be amazed that certain free concepts of the human mind actually fit the universe as perceived "out there," for the body/minds of humans are involved to some degree in the realization of both.

Through their body/minds and ecological relationships, and through the imaginative metaphors that these relationships inspire, humans create a human world reality, shaping it artistically. This is what Eli Bornstein called "naming" our experiences "from words to paint and stone, from poetry to architecture and music . . . bringing us into new consciousness about nature, new awareness of different levels of nature."[15]

What is important, it seems to me, is realization that the truth and the goodness of all our naming endeavors, artistic and scientific, are not purely relativistic nor are all variants equally valid. A non-human organic universe exists within which, and constrained by which, we fashion our experiential human universe. Progress is to get a better and better fit between the two, using all the skills of mind: intuition, imagination, common sense, and ethical sense, as well as a rationality that recognizes its dependencies on these other faculties, acknowledging its inability to disclose absolute truth.

The Importance of Metaphor

If the basic concepts and structures by which we know and feel are closely tied to our physical bodies, as examination of language suggests, why has the body been ignored? Lakoff and Johnson suggest several hypotheses foisted on humanity by objectivists:

Reason is believed to be abstract and transcendent, not tied to any bodily aspects of human understanding; The body introduces subjective elements irrelevant to the objective nature of meaning.

These traditional judgments are masculine and gender-biased. They rest on subtle, misleading metaphors: mind is male, body is female; mind is spiritual, body is material; mind is soul, changeless and eternal, body is changeable and corruptible; mind is good, body is bad or, in kinesthetic language, mind is up (ethereal) and body is down (mundane).

Clearly the metaphors by which we think and create can be uniting or separating. Lakoff and Johnson point out the demeaning effects of the metaphor, "labour is a resource," that reduces people to raw materials. Without care, the metaphors of the sciences and arts can also be divisive, disconnecting, separating. Let us recognize and acknowledge the existence of good metaphors and bad, and therefore good language and bad language, good science and bad science, good art and bad art, regardless of the excellence of the technical execution. Joining, not separating, is the standard of appraisal, and ecological discrimination is essential.

As an example, ask which is the better metaphor for "life,"

"organism," or "Earth?" Life, the unknown organizing force, has never been adequately defined. Yet organisms, things like us, have been taken to be life's equivalents and used as its metaphor—even though we know that organisms cannot exist without Earth, that they are born from and sustained by Earth.[16] Is not Earth a better metaphor for life than all the organic parts evolved within it? The implications of such a changed metaphor are many. For example, the misnamed "abiotic" environment that evolved all organisms and keeps them alive is the same Earth matrix from which we draw "raw materials" and "natural resources" with a payback of industrial pollutants.

Much more than a treasury of potential human wealth, Earth is the Mother of all aliveness.

To comprehend fully the centrality of Earth is to reduce our sense of separateness on many fronts, pointing new directions for progress in the twenty-first century.

References

[1] Wallace Matson, quoted in George Sessions, "Reinventing Nature, the End of Wilderness? A Response to William Cronon's *Uncommon Ground*" *The Trumpeter* Vol. 13 No.1 (Victoria: LightStar Press, 1996) 33-38.

[2] Schumacher, E.E., *A Guide for the Perplexed* (New York: Harper Colophon Books, 1978) 120-140.

[3] Ralston Saul, John, *The Doubter's Companion* (Toronto, London & New York: Penguin Books, 1995).

[4] Saul, 93.

[5] Spretnak, Charlene, *The Resurgence of the Real: Body, Nature and Place in a Hypermodern World* (New York: Addison-Wesley Publishing Company, Inc., 1997).

[6] Saul, 250.

[7] Holton, Gerald, *Einstein, History, and Other Passions: The Rebellion against Science at the End of the Twentieth Century* (New York: Addison-Wesley Publishing Company, 1996).

[8] Ziman, John, "What is Science?" *Philosophical Problems of Science and Technology* ed. Alex C. Michalos (Boston: Allyn & Bacon, Inc., 1974) 5-27.

[9] Capra, Fritjof, *The Web of Life: A New Scientific Understanding of Living Systems* (New York and Toronto: Anchor Books, Doubleday, 1996).

[10] Lakoff, George. and Mark Johnson, *Metaphors We Live By* (Chicago and London: The University of Chicago Press, 1980).

[11] Havel, Vaclav, quoted in Gerald Holton, *Einstein, History and Other Passions* (New York: Addison-Wesley Publishing Company, 1996) 35.

[12] Whitehead, A.N., "The Romantic Reaction" *Science and the Modern World: Lowell Lectures 1925* (New York: Macmillan, 1925 and 1948) 75-96.

[13] Johnson, Mark, *The Body in the Mind: The Bodily Basis of Meaning, Imagination and Reason* (Chicago and London: The University of Chicago Press, 1987).

[14] Lakoff, George, *Women, Fire and Dangerous Things: What Categories Reveal about the Mind* (Chicago and London: The University of Chicago Press, 1990).

[15] Bornstein, Eli "Dialogue on Art and Ecology," *The Structurist* 37/38 (Saskatoon: University of Saskatchewan, 1997/98).

[16] Rowe, J.S., "Earth as the Metaphor for Life." *BioScience* Vol. 48 No. 6 (1998) 428-429.

6. THE REVIEWER

REBELS AGAINST THE FUTURE: THE LUDDITES AND THEIR WAR ON THE INDUSTRIAL REVOLUTION; LESSONS FOR THE COMPUTER AGE. Kirkpatrick Sale. (Don Mills, Ontario and New York: Addison-Wesley Publishing Company, 1996); 320 pp. Reprinted with permissions from *The Structurist*, No. 35-36, 1995–1996. University of Saskatchewan, Saskatooon, pp 144–49. Edited for this collection.

This is a worthy addition to the literature of apprehension focused on big technology. Sale's *cri de coeur* resonates with the heartfelt alarm sounded in recent years by such as Ursula Franklin (*The Real World of Technology*), Neil Postman (*Technopoly*) and Jerry Mander (*In the Absence of the Sacred*). Their forebodings about the mechanized future into which technology is whirling us are catching public attention, though generating little response.

Perhaps convinced that the only believable parts of newspapers are the comics, the author begins with a prophetic quote from Daddy Warbucks addressing workers losing their jobs to automation: "Your real enemy—one you can neither fight nor reason with . . . it's not a 'who.' What you're up against is the Future." (Little Orphan Annie comic strip, 1982).

Few can doubt that, on present course, Daddy Warbucks' envisioned future is "Technopoly": the entire world appropriated for industrial use, computerized and robotized, rendering obsolete the majority of humans. Kirkpatrick Sale's reaction is an impassioned call to rebel against this repugnant future, to oppose the microchip second Industrial Revolution. His sympathetic historical sketch of the Luddites who resisted the first Industrial Revolution as it literally "gathered steam" includes an appraisal of their strengths and weaknesses, setting the stage for his analysis of today's machine bred problems along with a few hints as to the strategy that might overcome them.

Sale comments on the apathy of a public caught up in the contemporary firestorm of technological change. Everywhere people are insecure and worried. In the USA, 40 per cent of the labour force is now in "disposable" jobs with no prospects for steady work and advancement. Yet why so little concern with the deepening catastrophe of the second Industrial Revolution? The reason, he suggests, is that as a society we are ignorant about the past—we simply do not know the extent of the human and environmental traumas caused by the first Industrial Revolution. Stories told of the latter usually omit the catastrophes. If our sense of history were better we would be asking, "What will happen to the dispossessed this time around? What will be the consequences once our world, too, has been turned upside down? What will happen after the ecosystems and their species are destroyed? Once the line of ecological peril is crossed, what will the results be?" This, he says, is why the Luddites are important. Their story helps us understand the catastrophic consequences of a world suddenly transformed, degraded, dispirited, destroyed. Forewarned, perhaps we can do better the second time around.

The book's thesis is that a world dominated by the technology of industrial society is fundamentally more detrimental than beneficial to human happiness and survival. Of course, technology cannot be condemned wholesale. Some is benevolent (and I agree with the author that his book is in the "benign" category) but a great deal of it is malevolent to the user, to the community, to the culture, to the environment, to the future. What should be opposed, he says, is "Machinery hurtful to Commonality," a phrase of the Luddites that neatly summarizes their ethos. Note here an implicit equating of technology with machinery rather than with "all the cultural ways of doing things," the latter definition suggesting technology's important cultural/ideological context. In an unintended way, Sale's book is an exposure of the error of attacking "technology as machines," instead of chopping away at its ideological roots.

The Luddites went after the labour replacing machines, head on. The scene of the rebellion was the British heartland, the parts of Yorkshire, Lancashire, Derbyshire and Nottinghamshire whose principal cities are Leeds, Manchester and Nottingham. The same area was the home ground of Robin Hood victimized by an earlier industrial policy, that of clear-cutting the forest commons for the private-land grazing of sheep. As a result of that early enclosure, wool weaving became the key industry of the Midlands in eighteenth century England and woolen cloth its most important export. The livelihood of artisans—the weavers, combers and dressers of wool, as well as those in the later-developing cotton trades—was based on cottage industry, employing simple machines manageable by a family.

In the forty-five years from 1785–1830, British society underwent a transformation unequalled before or since, so drastic that "human nature and human need were made over." New, large, complex and polluting machines powered by coal and steam began to appear, housed in huge multi-story buildings, threatening the settled trades, creating unemployment, intruding on and destroying the ordered society of craft, custom and community. Thus technology based on the steam engine was imposed, causing the shift from an organic economy grounded in land and labour to a mechanical economy of coal, factory and foreign trade. Then as now, the machine dictated large-scale production, centralization, specialization, efficiency. It drastically reduced the value of personal skills, craftsmanship, aesthetic expression, social contacts, human autonomy, individual choice and especially human labour.

In the Midlands by 1833 steam power was calculated to be doing the work of 2.5 million people, almost equivalent to the labour force of the time. In the words of the prime apologist Andrew Ure, "It is in fact the constant aim and tendency of every improvement in machinery to supersede human labour altogether, or to diminish its cost by substituting the industry of women and children for that of men." No restrictions were placed on the employment of children starting at ages as young as four or five years, and by 1833 children and women made up four fifths of the labour force while males suffered massive unemployment. All who had jobs were slaves to the machine, working long inflexible hours—at least twelve and sometimes eighteen—behind locked doors, subject to a regimen of shop floor penalties, including physical punishment (strapping) for falling asleep on the job.

The ills attending destruction of cottage industry by the new wool and cotton machines were exacerbated by a parallel revolution in agricultural land use whereby more than six million acres of "commons" were privatized and enclosed. Traditional sustenance from small-scale farming and pasturing, gleaning and foraging, fishing and hunting, disappeared in a lifetime leaving the common folk helpless. The defining achievement of the Industrial Revolution was creation of a society in which people were given the choice between starvation and wage labour. And the wages offered reduced labourers to paupery. The suffering poor, "hollow-eyed urchins and bone-thin labourers" comprising at least a third of England's population, were supported in the most meagre fashion by Poor Law relief. Life expectancy in 1842 in Manchester-Leeds was forty-one for Jane Austen's "gentry" and eighteen for labourers. Yet, when in 1811 unemployed cotton weavers asked for relief, Parliament replied it would be improper to grant pecuniary aid, since that might destroy "the equilibrium of labour and employment."

Add to the human misery the devastation of the English landscape by factories belching smoke and poisoning the streams, and by the new agriculture exterminating wild plants and animals. "We war with Nature and, by our resistless engines, come off always victorious, and loaded with spoils," said Carlyle. "A world with nothing left to the spontaneous activity of Nature," said J.S. Mill, "with every foot of land brought into cultivation . . . and scarcely a place left where a wild shrub or flower could grow." Marx's patron, Engels, told the story of remarking to a manufacturer in Manchester that he had never seen so ill-built and filthy a city. "The man listened quietly to the end, and said at the corner where we parted: 'And yet there is a great deal of money made here; good morning, sir.'" Sale remarks that nothing could long stay healthy or alive, save only the indomitable spirit of avarice.

The threat to livelihood and lives of the labourers brought on widespread smashing of industrial machines, mostly in the period 1811–1813. The perpetrators—in some instances women as well as men—called themselves Luddites, taking their name from a legendary Ned Ludd whom they promoted, in the pseudonymous letters sent out under his name, to the rank of "General" or "King." According to the fable this young man, apprenticed to a knitter near Leicester, was so reluctant to work at the knitting frame that his master got a magistrate to order him whipped, whereupon Ned took a hammer and demolished the frame, an act of such renown that thereafter when any machine was damaged people would say, "Ned Ludd's been here." In the industrial age, Ludd was a better symbol of revolt than Robin Hood who robbed the rich but did nothing to deprive them of their destructive toys.

> Chant no more your old rhymes about bold Robin Hood,
> His feats I but little admire.
> I will sing the Achievements of General Ludd,
> Now the Hero of Nottinghamshire.

After fifteen tempestuous months of sabotage the battle to ban "Machinery hurtful to Commonality" was lost. Capital triumphed over labour. The winners—aristocratic politicians, manufacturers and factory owners—set the subsequent terms of discourse. And so the word "Luddite" became pejorative, calling up images of machine-wreckers, ignorant or insane workers bent on obstructing industrial advancement. Today, opponents of new technology are again labelled "Luddites," symbolized thereby as unenlightened losers worthy of disparagement and easily ignored.

The Luddite threat to the established order (both real and exaggerated) called forth the greatest spasm of repression ever used by the British government against domestic dissent. Spies, special constables, volunteer militias and posses, midnight raids, hanging judges, harsh punishments and a force of soldiers greater than Wellington had taken to fight Napoleon four years earlier were the order of the day. Luddites were understood as representing a serious threat to industrial progress—they were rebels against a future assigned to them by those who controlled capital. These early capitalists could do anything they wished, encouraged and protected by government and king, unrestrained by laws, customs or ethics. The challenge posed by the Luddites, although they did not fully understand it, was to the economic ideology of the time, calling into question, on grounds of justice and fairness, the legitimacy of the principles of unrestrained profit, competition, innovation. The architects and beneficiaries of the new industrialism were clearer sighted; they knew it was imperative to subdue the challenge, to deny and expunge the premises of ancient rights and traditional mores in order to make the labour force malleable.

In less than two years the rebellion faded. At its core Luddism was a howl of protest and defiance. When it was met with hostility and repression it hardly knew what to do, how to continue, where to move. It ebbed after making its point that the misery, pain, humiliation and displacement accompanying the "progress" of industrial capitalism was hurtful and odious and demanded resistance and rebellion. Deterrence doubtless played a part in Luddism's demise: possibly three dozen Luddites were killed during factory raids, twenty-four were hanged, the same number imprisoned, and fifty-one sentenced to Britain's gulag, Australia.

As the information based second Industrial Revolution sweeps all before it, Sale warns us to heed the Luddite admonitions. Beware the technological juggernaut, reckon the terrible costs, understand the worlds being lost in the world being gained, reflect on the price of the machine and its systems on your life, pay attention to the natural world and its increasing destruction, resist the seductive catastrophe of industrialism. The message is urgent in this era of postmodern, transnational, "third-wave" capitalism characterized by new high-tech industrialism of ever more complex machines—computers, robots, satellites, artificial intelligence, biotechnology—served by remote institutions, including universities, and cheered on by the servile governments of "developed" nation states. As before, the future painted by technophiles is brimful of specious promises: let the warning trumpets sound! Soon to be here: the Information Superhighway, the Fully Automated Battlefield!

In retrospect, many of today's trends were initiated during the period of Luddism, for example, the conservative "reformism" of labour that resigned itself to an industrial future when its revolutionary goals were frustrated. Radicalism died. Labour unions are no longer rebels against the future but participants in it, their primary goal to get a bigger slice of the pie. Another trend, steadily strengthening, is the alliance forged between industry and government that today sees the two going off together, hand in hand, to drum up export trade in the far corners of the Earth while social programs are slashed at home. The true business of government is no longer the welfare of its citizens—provision for "the common good"—but the welfare of its corporations and business people. To this end taxes must be reduced and social services trimmed.

The alliance of bureaucrat and plutocrat also helped forge the idea of "class," a useful construct for social stratification and manipulation. The concepts of "ruling/upper class" and "working/lower class" make the division between rich and poor seem natural, as if it were God's will to separate the worthy from the relatively worthless. This in turn has confirmed the permanent breach between worker and master, the latter relieved of responsibility for the former. And as a final "downer," governments learned from the Ludditic experience the techniques of quelling internal rebellion—the use of overwhelming army power. Spies, bribes, arrests and hangings were effective in breaking worker resistance. At least in the ranks of those who mattered, such action did *not* call forth the outrage and denunciation expected in a nation supposedly devoted to individual rights and legal niceties. Notably ignored, because they championed a better deal for the poor, were the good poets Blake, Wordsworth, Shelley and Byron, the latter with a seat in the House of Lords yet penning a verse that ended, "Down with all kings but King Ludd!"

What is the positive legacy of Luddism today? Sale asserts that it brought into focus "the Machinery Question." In the words of one post-Luddite pamphleteer, "Unrestrained machinery demoralizes society" and "substitutes idleness for industry, want for competence, immorality for virtue . . . and unless restrained, will, ere long, involve this country in every horror and calamity attending the bursting of all the bonds that hold society together." When the development and application of a nation's technology is unquestioningly put in the hands of scientists, technocrats and business heads, the maintenance of social bonds sinks to the bottom on the list of political priorities.

Corporations make the argument, "What's good for General Motors is good for America" or, in simplified form, "The more technology, the more jobs." In theory, technology is supposed to increase jobs by making factories more productive, turning out more for less, stimulating a demand that should eventually spin off into more shops, more employment. Something like this *did* happen in Britain, but only after several generations of massive unemployment during which those unfortunates superseded by machines lived in hopeless poverty. What eventually created more jobs in England, mostly at the expense of people elsewhere, was government investment in colonialism, in war, and in less virulent forms of spending at home. By itself, technology has always abolished work. Consider today that "service jobs"—the hoped-for providers of employment as automated resource-based industries "downsize" and shed themselves of employees—are next in line to be mechanized. The role of humans in all kinds of production is bound to diminish, said the Nobel economist Wassily Leontief. Yet still the worry is expressed that, as populations age, insufficient youngsters will be on hand to fill "the workers" niche and by their taxes support the retirees!

Sale devotes a chapter to the parallels between the first Industrial Revolution of Luddite times and the "post-modern" second Industrial Revolution now upon us.

> It is remarkable how the new period resembles the old in many little ways: the early 19th century was a period of vulgar theatre, elephantine buildings, public obsession with murders and executions, increasing fear of street crime, great enthusiasm for boxing and other violent spectator sports, and passions for running, ballooning, and gambling of all kinds, including lotteries. But it is remarkable also that the large characteristics that served earlier to define the first Industrial Revolution can be applied . . . to the second as well."

According to Sale, the six "large characteristics" common to both Industrial Revolutions are:

1. the imposition of complex technology without democratic assessment;
2. the destruction of past ways of working and living justified as technological "progress;"

3. the manufacture of needs (wants) that accompanies urbanization and militarization abetted by advertising (especially today via television);
4. the support of all forms of high-tech by governments;
5. the ordeal of labour as jobs either eliminated or exported;
6. the destruction of Nature as Earth's wealth is converted to throw-away human wealth.

This last merits special notice:

> It is characteristic of industrialism . . . to make swift and thorough use of Nature's stored-up treasures and its living organisms, called "resources," without regard to the stability or sustainability of the world that provides them—a process ratified by such industrial ideologies as *humanism* which gives us the right, *materialism* which gives us the reason, and *rationalism* (science/technology) which gives us the method.

The three pedestals of Western culture—humanism, materialism, rationalism—underpin the second Industrial Revolution. How shall these pedestals be chipped away to bring "down to Earth" the current technology that is simultaneously rendering people obsolete while poisoning and eating up the Ecosphere? This key question, unfortunately, is peripheral to the theme of Sale's book. He makes brief reference to Chellis Glendinning, a New Mexico psychologist, who in 1990 proposed a "program for the future" that envisioned the dismantling of computer technologies as well as nuclear, chemical, biogenetic, electromagnetic and television technologies. She published "Notes Toward a neo-Luddite Manifesto," an attempt to give legitimacy to those who in one way or another are troubled by and resistant to the technology of the second Industrial Revolution.

"Neo-Luddites have the courage to gaze at the full catastrophe of our century," she began, which is that "the technologies created and disseminated by modern Western societies are out of control and desecrating the fragile fabric of life on Earth." Arguing that effective resistance requires new ways of thinking, and the creation of a new worldview, she advanced three basic principles of neo-Luddism:

1. Opposition to technologies that emanate from a worldview that

sees *rationality* as the key to human potential, *material acquisition* as the key to human fulfillment, and *technological development* as the key to social progress.

2. Recognition that since all technologies are political, the technologies created by mass technological society, far from being "neutral tools that can be used for good or evil" are inevitably those that serve the perpetuation of that society and its goals of efficiency, production, marketing, and profits.
3. Establishment of a critique of technology by fully examining its sociological context, economic ramifications, and political meanings . . . from the perspective not only of human use but of its impact on other living beings, natural systems, and the environment.

As with all worthy manifestos, she ended with a call for action—we have nothing to lose except a way of living that leads to the destruction of all life; we have a world to gain.

Is the idea of such a movement far-fetched in the Land of Technophilia? Sale thinks not, because even in societies that have succumbed to the new technologies an undercurrent of disquiet and fear exists, fed by science's awesome achievements at Hiroshima and in German death camps, in Bhopal and Chernobyl, in PCBs and Love Canal, in global warming and destruction of the ozone layer. Yet the argument is weakened by the fact that even with these horrors before us, technology in all its varied forms marches on. Some might point to a neo-Luddite victory in pushing nuclear energy offstage, but still it is far from banished; it waits in the wings to be called out again at an opportune time by the military, or by governments in the next decade when fossil hydrocarbons run low.

Sale recognizes the difficulties in the way of technology-refusniks when he says, "Technology always has consequences, far reaching and *usually more so than anyone can predict*" (emphasis added). He tells the story of IBM bringing together a learned group to discuss the implications of the computer for American society. After a week of discussion the experts threw up their hands, saying they could not possibly foretell the range of impacts. As one participant pointed out, if Henry Ford in 1910 had assembled the best minds to ponder the implications of the automobile in America, they could not possibly have predicted the automobile's personal, familial, social, architectural, cultural, industrial, economic and environmental effects—and the computer is far more versatile than the car.

It seems to me that society embraces each new technology because it promises *specific* contributions to comfort, health, convenience or efficiency, while its deleterious effects are *non-specific*, hidden in the future, unforeseen and in advance unprovable. Baboon bone-marrow cells are infused into an individual as a possible cure for AIDS, despite the possibility that equally virulent viruses could be transferred from baboons into the human population. Note in this example that the expected benefits are specific, practical, and readily appreciated; the drawbacks, if any, are distant, diffuse, slippery and eminently arguable. So long as technology is judged solely in terms of its contribution to *people*, nothing can stop it. Were we to judge technology *first* by its effects on the health of Earth, and only then in terms of its convenience for humans, solid reasons for technology refusal would more readily appear. By this criterion, huge airplanes would immediately disappear from the skies, and railways not trucks would be the favoured means of transporting goods.

With Glendinning's manifesto in mind, Sale asks what lessons from the first Luddites might help today's citizens be more effective "rebels against the future." Primarily the truth must be accepted that technologies are never neutral and some are harmful. To believe otherwise is to adopt, in Marshall McLuhan's immortal words, "the numb stance of the technological idiot." Every technology carries an inevitable logic, bearing the values and purposes of the socio-economic system that produced it. Guns and their use express the values of a violent culture. In a violent society, beating swords into ploughshares diverts war against humans to war against Earth's soils.

Wendell Berry suggested that each technology, such as a new tool, be judged by these criteria: Is it cheaper, smaller and better than the tool it replaces? Does it use less (renewable) energy? Is it repairable? Does it come from a small local shop? Does it replace or disrupt anything good that already exists (including family and community relationships)? Sale sagely adds that "family and community" should be understood to embrace all other species, plants and animals alike, and the living ecosystems on which they depend, and that they be considered (as the aborigines have expressed it) with the interests of the next seven generations in mind. These ideas tie in with the maxim of art critic Herbert Read:

> Only a people serving an apprenticeship to Nature can be

> trusted with machines. Only such people will so contrive and control those machines that their products are an enhancement of biological needs, and not a denial of them.

Few politicians or CEOs have served "an apprenticeship to Nature." Hence political and economic leaders do not know that, in the same way as the human species is one dependent component of nature, so the human economy is a dependent subset of Earth's economy. The nation's god, "the economy," has been developed into a totally parasitic, destructive part of the world rather than into Earth's friendly symbiont. Technology tends to distance us from Nature, an idea neatly captured by the definition, "Technology means so arranging the world that one never has to experience it directly." Television and the computer screen are perfect examples. But the ecological bond between human institutions and Earth ought always to be foremost in mind. In the words of Jacques Ellul, "The artificial world is radically different from the natural world," with "different imperatives, different directives, and different laws" such that "it destroys, eliminates or subordinates the natural world."

Unless inhabitants of the technosphere recognize and re-establish connectedness with the Ecosphere, they will destroy their source and eventually themselves.

Neo-Luddites must understand that industrialism in its rapid growth phase is a cataclysmic process making for an uncertain future, that the nation-state will always come to the aid and defence of industrialism, that resisters must force "the machine question" and the viability of industrial society into public consciousness and debate. The most important lesson, Sale says, is that resistance to the industrial system must be embedded in an analysis—an ideology, perhaps—that is morally informed, carefully articulated and widely shared. "All the elements of such an analysis . . . are in existence, scattered and still needing refinement, perhaps, but there—in Mumford and Schumacher and Wendell Berry and Jerry Mander and the Chellis Glendinning manifesto; in the writings of the Earth-firsters and the bioregionalists and deep ecologists. . . ." And his closing statement, "It is now the task of the neo-Luddites, armed with the past, to prepare, to preserve, and to provide that body of lore, that inspiration, for such future generations as may be."

The hoped for new worldview has yet to crystallize. Its elements may be in existence as Sale says, but scattered around. The inspiring ideology, the "body of lore" that will undercut the pillars of anthropocentric humanism, materialism and the narrow rationalism of science/technology—throwing the shards in the gears of the machine-world throbbing around us—is still a vague and distant hope. Could the first step be a rejection of human narcissism, an emphasis on the more-than-human? Sale does not say. In a way his book is a throwback to nineteenth century Luddism, focused primarily on "the Machine Question," beyond that unsure of goals and directions. His is a necessary, stentorian call to attention, not the marching directives of the day.

HOW DEEP MUST THE CHANGE BE?

Reviews of *In the Absence of the Sacred: The Failure of Technology & the Survival of the Indian Nations*, Jerry Mander, (San Francisco: Sierra Club Books, 1991); and *The Dream of the Earth*, Thomas Berry, (San Francisco: Sierra Club Books, 1990). Reprinted with permission from *The Structurist*, No. 31/31, 1991–1992. University of Saskatchewan, Saskatoon, pp 144–47. Edited for this collection.

Hundreds of books have been written on the failure of contemporary society to come to grips with the problems of the Earth-environment that it is degrading. A general conclusion has been that the good-hearted response of many—recycling, re-using, not littering—does not quite meet the challenge. Rather, changes must be made in human attitudes to surrounding Nature. Attitudes reflect people's values and therefore the call is really for changes in what the general public perceives as most important. Beyond humanity the outer world must be valued in new vital ways if the degradation is to be stopped.

How fundamental and deep-seated are the needed changes in values and attitudes? The Norwegian philosopher Arne Naess suggested that the reigning superficial view, with its trust in science/technology to cure environmental woes, be called "shallow environmentalism." In contrast to this prevalent human-centred faith Naess advocated "deep ecology" based on recognizing the intrinsic values of all organic life-forms on Earth. An even deeper ecology values Nature as Earth, its land and water ecosystems, including all their dependent life-forms of which humanity is one. I have chosen two books to illustrate the shallows and the depths of environmental thinking.

Both Jerry Mander and Thomas Berry examine the cultural basis of current social and environmental problems, meaning by "cultural basis" the taken-for-granted values and activities of Western civilization. Mander's title, *In the Absence of the Sacred*, is slightly misleading because little in the book is devoted to "the Sacred" and its absence. What rouses his ire is society's uncritical acceptance of new technology as "progress," not asking who benefits, never questioning the corporation-science-government-university network that sponsors every kind of novel technology, not calling to account the self-appointed technical elitists who grant themselves the right to follow any lead that promises a marketable product and a money reward. Mander joins Neil Postman in putting technology under the microscope and exposing its dangers. Today's society is Technotopia, unexamined, uncriticized, lurching along the road to ever-greater socio-environmental disasters.

As in his earlier book, *Four Arguments for the Elimination of Television*, Mander takes TV as the exemplar of all that is bad in unscrutinized technology. Since 1945 nearly ten thousand books have been written on TV, but none (his book excepted) has argued that society would be better off without it. According to him, TV homogenizes cultures, makes people more passive, less able to deal with complexity, less able to read, dumber, and with less understanding of world events. TV is isolating and addictive, a hypnotizing instrument of mass brain-washing. The average "hooked" North American takes a five-hour "hit" every day. TV trains children, he says, to expect speed and excitement and kaleidoscopic change as life's norm; it is a drug, training for harder drugs. TV redesigns us to be compatible with the passive consumptive future that big governments and big corporations want. Most damning, the TV medium invites the appearance of actors as gurus and political leaders.

Similar criticisms are levelled at computers and satellites, at cars and the products of genetic engineering. Who benefits most from these gadgets and techniques? The military and transnational corporations. Seventy-five per cent of commercial TV time is paid for by the one hundred largest corporations who annually pump into the average head twenty-one thousand commercials, each saying "Buy something, and do it now!" And corporations—machine-like bodies guided by efficiency and greed in the voracious pursuit of profits—are both purveyors and creatures of technology. The phrase "good corporate citizen" is an oxymoron, for corporations *must* subordinate social and environmental responsibility to their legally sanctified bottom lines: growth and exploitation. Profit comes from transmogrifying raw materials into saleable forms and so, as corporations grow, nature is reorganized and degraded at an accelerating rate. The more human nature and non-human nature are exploited, inside our bodies and outside them, the greater the corporate profit. Small wonder that corporations propagandize for the technology that is their life blood, and co-opt the education system to that end.

Few people question technology, says Mander, because proponents always present best case scenarios: technological innovation will promote democracy, freedom, leisure, enjoyment, a better life. The flash and power of new machines is an additional seduction, especially for men. Further, technology is mistakenly assumed to be neutral and therefore its ideological bias toward exploitation, alienation, centralization and top-down control escapes detection. Consider the "home" computer versus the computer-satellite systems of large corporations and note the tendency to judge technology on a narrow personal utilitarian basis rather than by its overall political, social, spiritual, environmental and economic effects upon the world.

Technology is an expression of human thoughts about how to do things. Immersion in technology is immersion in the products of minds oriented to satisfying human desires, some good and some bad. For example, surrounded by machines whose goal is efficiency, we begin to respond in mechanical ways to this unconsciously accepted reality. To question technology is to question thought itself, our everyday lifestyle, our sanity.

Given that we exist within the proliferating technosphere—a thought-produced cocoon that reflexively remakes our minds in a frightening infinite regress—and that it is responsible for the worsening human and environmental conditions of this planet, is there any way out? Deep criticism of technology is part of the answer, writes Mander in recommending a "guilty until proved innocent" stance, but more, we need a paradigm shift, a new worldview that goes beyond *human* liberation. Providentially, a broader vision is embodied in the spirituality of the American Native people and their traditional relationship to the Earth.

Mander argues that expecting technology to live up to its own advertising, liberating itself from self-created problems, is romantic and unrealistic. The only people who are clear-minded on this point are the Native people simply because they have kept alive their roots in an older, alternative, nature-based philosophy that has proven effective for tens of thousands of years. It has nurtured dimensions of knowledge we used to have—Earth-wise dimensions now forgotten and hidden from us. Native societies, not our own, hold the key to future survival. Look at them, says Mander. Their strength is fed by the knowledge that what they are doing is rooted in the Earth and deserves to succeed. With humour and kindliness they fight on behalf of their values and for their descendants. Therefore, learn from them, get involved, don't despair—changes in Eastern Europe show how rapidly the world can be altered and how quickly paradigms can change—and do not be self-conscious about speaking of Mother Earth. Mander's optimistic message is learn, reason, understand and act. Note that the paradigm change to which Mander refers was from one kind of plundering, inefficient economic system (communism) to a more efficient one (capitalism). That kind of shift comes easily to humans.

The Dream of the Earth champions a deeper, more fundamental view, arguing that escape from technology's flashy lights (and from their inevitable burn-out) depends on a sympathetic, reverent, perception of the star-filled cosmos in both a literal (phenomenal) and mystical (noumenal) sense. Inspired by a larger-than-human vision of the creative universe, eco-theologian Thomas Berry has written a thoughtful text on the physical and spiritual aspects of human ecology. The roots of the human predicament lie within our minds but that, he says, does not make them opaque to intelligence. Their cultural source—the species-centredness of *Homo sapiens*—is obvious. Deliverance from this confinement will come from a renewed sensitivity to everything that anthropocentrism has shut out. We must strive for a wide and deep communion, broadened to embrace the *whole Earth*.

Every year ten thousand species are extinguished—"ten thousand divine voices stilled." Aggressive industrialization is destroying the world, laying waste forests and grasslands, eroding soils, poisoning air, water and soil, and all because of mistaken theories, beliefs and values. Chief of these are human-centred myths, a religious tradition emphasizing salvation from Earthly existence, a millennial vision energizing technologic "progress," pathological patriarchy in its various forms, a way of knowing (science) that eschews values. All contribute to a psychological people/Earth split that is foreign to the Aboriginal communities of the Americas.

We need a new mythic vision, a new story of where we came from and who we are, says Berry. Though science is out of touch with its creative spiritual roots, it nevertheless has provided a magnificent fifteen-billion-year revelation of dynamic change in a universe whose latest diversification has brought one small part of itself to self-consciousness. When humans celebrate their source the mysterious creative universe is celebrating itself.

Why so little celebration? Because long ago the Western religious tradition turned away from the wonders of existence on this beautiful planet and sought instead salvation from it. Concern with the duality Man/God, rather than with the trinity divine/Nature/humanity, is the outcome of a theology of original sin and suffering, of belief that a primordial human fault broke the harmony of the universe, necessitating the formation of a redemptive rather than a celebratory community. Also the dogma of a transcendent God denied the natural world as the meeting place of the divine and the human, leading to the idea of Nature as merely an external object. By insisting that the human is a spiritual being with an eternal destiny beyond other members of the created world, by elevating the human, we have alienated ourselves from the Earth. Yet our human fulfillment, Berry says, is not to be found in isolated grandeur but in intimacy with the larger dimensions of our being—with the Earth community to which we are tied physically and spiritually.

Salvation theology—strong on personal sins, weak on abuse of the natural world—carries a millennial vision of a re-made world of peace and plenty, transformed into an idea of endless material progress through science and technology that has inspired and energized in sequence the Age of Enlightenment, the democratic age, the nation state, the classless society and now the capitalist-consumer society. "Progress" has become mindless technology, devastating the planet. Its crowning glories are cities, chemical agriculture, robotized factories, nuclear weaponry, Disneyland and the West Edmonton Mall. The visible signs of patriarchy's leadership are rich wonderworld designed inside the industrial bubble, and outside it, trashed wasteworld.

Perhaps, Berry writes, the die was cast in Eurasia about 5500 years ago when the Aryan invasion ended matricentric civilization and initiated virulent patriarchy, "the archetypal pattern of oppressive governance by men with little regard for women, for the more significant human values, and for the destiny of the Earth." Religion was corrupted, nature rituals weakened and the archetypal world of the unconscious denied. Four outstanding expressions of patriarchy are the classical empires, the ecclesiastical establishment, the nation state and the modern corporation. All have shared or today share a vision of both human and natural worlds as arenas of strife for control, all have subjugated women, sought unrestricted power, and co-opted education.

Both the nation and the corporation devote much propaganda to the proposition that their own protection is a sacred duty. With support of the nations, corporations continually increase their skills in plundering nature, but they "have not the slightest idea of how to establish a mutually enhancing mode of human presence upon the earth." Berry looks to the unique spiritual tradition of the Native people of the North American continent and to the feminist anti-patriarchal movement, supported by the ecological movement, to establish the basis for a new historical period whose norms of reality and value are Nature-centred not human-centred, a post-patriarchy in which democracy is replaced by "biocracy." To my mind, an "ecocracy"—Earth-centred rather than organism-centred—would better answer what is needed.

One of the practical recommendations for "reinhabiting" the Earth in a safe and sane way is to follow the lead of bioregionalism (better ecoregionalism) whose primary ecological goal is to maintain the health and integrity of natural regions. In bioregionalism, people are encouraged to join the Earth-community as participating members, to foster the progress and prosperity of the bioregions to which, without realizing it, they belong. This is a "forward to Nature" call, an invitation to renew ties with the Earth, a summons to recognize people as cooperating parts in a larger whole more important than they. To find our way in the future we can no longer depend on cultural tradition but must trust the ancient nature whose sagacity and acumen is in our genes, in our "instincts," for we have wandered far from the Earth-wisdom that used to be ours.

These books make an interesting duo, expressing faiths whose foundations lie successively deeper in the human psyche. Mander believes that reason can extricate humanity from the technological new world order now threatening to engulf it, while Berry believes that Earth-spirituality, orienting humanity to its ecological/cosmological context, can bring about an artistic resolution of social and environmental antagonisms. Both authors find a source of hope in Native cultures whose animistic spirituality, surviving missionary efforts to eradicate it, offers, they say, a saving alternative to the prevailing Western mode.

Neither author discusses the feasibility of popularizing "alter/Native" Earth-animisms. What are the possibilities of grafting Aboriginal values on Western secular society? Would the norms of foraging and early agricultural societies make sense in our corporation-dominated, industrial society? Mander seems to think so, and Berry would effect the same transfer of Aboriginal values to the Western religious base. Both strike me as naïve hopes. A more fundamental change in the way the human race views itself in relation to Earth is the "paradigm shift" most needed. Berry comes close to expressing it, having looked at the biological, ecological, and evolutionary evidence that is building. Once we begin to feel ourselves Earthlings, all appropriate values—both the rational and the spiritual—will reveal themselves.

THE ECOLOGY OF EDEN

Review of *The Ecology of Eden*, Evan Eisenberg, (Toronto: Random House of Canada, 1998); 632 pp. Reprinted with permission from *The Structurist*, No. 39/40, 1999–2000. University of Saskatchewan, Saskatoon, pp 77–85. Edited for this collection.

First came the doom-sayers proclaiming the death of Nature and then, in reaction, the technological optimists pooh-poohing prophesies of imminent environmental disaster. Inevitably the extremes invited a somewhere-in-between view, and this Evan Eisenberg has provided in *The Ecology of Eden*. It is a critique and synthesis of the contrasting beliefs of those who think humans are controlling the planet too much, and those who believe environmental problems signal insufficient control. I judge the book important because, in a very modern way, it illuminates the ancient question—How can humanity live well on this planet Earth?

A cosmopolitan New Yorker, Eisenberg represents the new breed of intellectual: a humanist well versed in the biological and ecological sciences. Educated in philosophy/classics at Harvard and Princeton, in biology at the University of Massachusetts, and an established writer on nature, culture and technology, he has been a music columnist for *The Nation*, a synagogue cantor, and a gardener for the New York City Parks Department. Such a background guarantees intelligent comment on ancient myths and biblical literature, on wilderness and cities, on gardens and jazz. Further, Eisenberg is at home in the language and uses it exuberantly. He knows a haruspex from a harpy.

His wide-ranging scholarship, confirmed in a copious 435 pages of witty text, is further emphasized by a book-sized compendium (155 pages) of references in Bibliography and Notes, the latter described in part as "Scholarly quibbles, qualifications, equivocations, and pygocryptions . . . tidbits too tangential to put in but too juicy to leave out." These appended remarks are not trivial; they must be studied to get the most from the text.

In one of his essays Aldous Huxley advised vigilance where facile writers are concerned because, he said, they can so easily lead those of us with weaker minds astray. Sailing along early in the book this reader's sonar sounded alarms when shallows and partly submerged agendas were detected. A few examples: on the subject of the relationship between people and nature those at the two extremes are dubbed "planet Managers" and "planet fetishers," and of the two only the latter is clearly pejorative. Deep ecologists are placed in the "fetishers" camp and scathingly attacked on the basis of a few misguided statements by disciples, without reference to the peaceful philosophy of the founding father, Arne Naess. To Eisenberg, "ecology" means conflict, and an oft-repeated phrase is "conquering the world," indicating that he accepts the idea of a competition-ruled, non-cooperative biosphere. Frequently he uses "fully human" in reference to today's civilized people, implying that we know the "fully human" goal and are almost there. Again, the voice throughout is that of a defender of the Hebraic faith. All writers have their axioms, but in a significant primer such as this is, their early exposure would be appreciated.[1]

The book's central proposition is that humanity needs to nurture, geographically and psychically, both the wild and the civilized, symbolized by the Mountain and the man-made Tower. Wild nature is the source of life; it is Eden as exemplified in wilderness areas. The civilized is the source of culture, exemplified by large cities, their artifacts and activities. The inscapes of our minds match this duality: we are both wild and cultured, partly animal/instinctive and partly human/self-conscious. Nature and nurture are essential elements of our being. Dreams of Arcadia, half wild and half civilized, express the human desire to have the best of both worlds.

The "planet fetishers," yearning to live in a wild world that excludes civilization, are as romantically unbalanced as the "planet managers" who are preparing, with satellites and computers, to vanquish the wild and make all Earth an instrumented space-ship. The wish for a simple back-to-nature existence is as unwise as the wish to be thoroughly citified. Eisenberg's advice: settle for a continuously improvised piece of both. By avoiding extremes people can have their urban cake and Eden too.

A merit of the book is its reminder, chapter after chapter, that humanity's aptitude for warring on wilderness—converting nature's wealth into human wealth without thought of moderation and sustainability—is as ancient as history. The only difference between our civilization and Sumerian civilization six thousand years ago is that today, with more people and more power, we can do the job faster. Thus the tension between civilization and wilderness is a theme as old as the recent post-glacial epoch—the Holocene—in which agriculture began. "Let's Call the Holocene Off" is one of Eisenberg's tantalizing headings, and ancient stories suggest our ancestors might have applauded it. Their myths tell of humanity's major mistakes—the attempts to live wholly in the wilderness or wholly in the city.

Best known in the West is the Genesis fable, locating Adam and Eve as the first foragers on God's mountain wilderness—the Garden of

Eden—source of the four great rivers of the world and fertile rooting for the Tree of Knowledge and the Tree of Life. Eisenberg notes Adam's dilemma when "put . . . into the garden of Eden to dress it and to keep it" (Genesis 2:15). How can he work it (dress it) and at the same time protect it (keep it)? Having eaten from the Tree of Knowledge, Eve and Adam must have realized that God was setting them up. The solution—the first example of land-use zoning—was to depart from the Garden of Eden, "keep" it as a wilderness preserve, and "dress" the valley bottom where the Edenic rivers drop fertile alluvium, the best agricultural soils. People cannot inhabit paradise without changing it to something non-paradisal. Therefore, leave God the mountain wilderness from which the healing waters flow; grow wheat, and cities, and civilization, on the flood-plains down below.

Evidently the "fall of man" was an exchange of the easy fruit-picking life on the uplands for the hard work of farming on the lowlands. Women invented agriculture that brought about their own demotion, for the "forbidden fruit" was wheat not apples. Ten thousand years ago humans turned away from the wandering foraging lifestyle, opting for the sedentary tending of crops with its promise of porridge every morning. Wealth in the form of storable and transportable grain gave rise to cities, kings, bureaucracies, armies, war. Men strong-armed their way to dominance. Patriarchal society, cities, and religions of the Book, are a legacy of early till-agriculture.

The Garden of Eden myth symbolizes a fancied lost Golden Age when people lived simply and innocently, akin to wild animals, before they began excoriating Earth with the plough. In truth, the Mountain "where God planted a garden," with its "Tree of Life," is the source of Earth's vitality. It is a World Pole, an *axis mundi* to which our wild nature is attuned. Essential for a healthy environment and the health of our lives, it tugs at our instincts, drawing us away from city, civilization, culture.

Eisenberg would restore and preserve Eden in the mountains, make great areas of wilderness an inviolable part of the planet marked "out of bounds" to humans, a worthy proposal but already foreclosed for most parts of Earth. Although we cannot live as wild beings, we must protect the uncivilized World Pole on whose ecological services humans and other organisms depend. Here note, in the interests of proper use of language, that Eisenberg's purposes for "preserving Eden" are as human-serving and homocentric as the purposes of the city. The less catchy title of the book should have been "Human Ecology and the Uses of Wild and City."

People also need the second life source: the civilized World Pole symbolized by the manufactured Tower. The original is the terraced ziggurat of Mesopotamia, built by the Sumerians and Babylonians on the broad floodplains of the Euphrates river. There in a man-made paradise of cereal crops, fruit trees, and gardens, cities of baked mud brick arose, each with a sacred centre and a stepped pyramid often crowned by a temple. Apparently the Sumerians were incapable of imagining a world without agriculture, irrigation, and cities, and they attributed the genesis of all three to the gods. The ziggurat, the city-built Tower, was their World Pole: a sanctioned *axis mundi*. It fostered culture, technology, and the comforts and stimulations of civilization, answering another set of innate human needs.

Southern Mesopotamians, an urban folk dependent on irrigated agriculture, considered pastoral life in the hills unacceptable. Humans were subhuman "until the gods shoved hoes into their hands." Perhaps they can be excused for a hubris akin to our own, for they were the Edisons of their time—inventors of wheeled vehicles, yokes and harnesses, animal-drawn ploughs, sailboats, metal working, the potter's wheel, the arch, the vault, the dome, surveying, mapping, a rough-and-ready mathematics, and cuneiform writing—all fruits of the city. In parentheses, the parable of the collapse of the super-ziggurat—the Tower of Babel—was written by northern Semitic scribes distrustful of cities and contemptuous of skyscrapers as stairways to paradise. Taking a leaf from Eisenberg, they would have endorsed the modern warning that ziggurats are injurious to human health.

The Sumerians and Babylonians had inklings of their dependence on forested mountain wilderness, the vital source of their wood and irrigation water. The first recorded epic of the Bronze Age tells of a king, Gilgamesh, who ruled the city of Uruk in southern Mesopotamia five or six thousands years ago. To build and embellish Uruk, he and his "noble savage" companion Enkidu logged the sacred cedar forest in the mountainous "land of the living" adjacent to the fertile crescent. To accomplish this archetypal clear-cut the two heroes first had to do away with the forest guardian, the storm-god Huwawa, by axing his neck. Then the Tower-men skinned the mountain of its cedars. Eisenberg comments:

> This is a turning point for the Western mind. Nature offers its services: flood control, pest control, climate control, gene

> banking, renewable resources without end. Of all these boons there is no better symbol or dispenser . . . than the forest. In brief: energy, information, and the two combined in ten thousand kinds of work done for us absolutely free of charge—"on the house," we might say, with a nod to the root of "ecology." An offer we couldn't refuse, you might think. And what is our answer? Off with its head! Nature is decapitated, its energy cut off from the information that once directed it.

An illustration of Eisenberg's acumen is this added comment:

> Noteworthy is the fact that it is Enkidu who is eagerest to finish the monster (Huwawa) off. No zealot like a convert—and no pillager of nature like the "natural man" who has had a taste of Western technology and lifestyle. From ancient times to the present, one of the basic tools of imperialism has been to make subject people its accomplices in the rape of their homelands. With any luck, they will be the ones who suffer the consequences. In the epic, indeed, it is Enkidu whose death is brought about by the train of events that the conquest of the cedar Forest set in motion.

The Gilgamesh story of a sacred mountain forest felled to build a great city contrasts with the Eden story of a life-giving paradisal mountain where God walks in the cool of the day. In the making of civilization, Eden was unmade. According to Eisenberg, "The love of nature and the urge to master nature have always, I am sure, been basic to the human mind."

The dilemma, Eisenberg writes, is the need, deep in the ancestral body/mind, for wild nature, for Eden, and not just for city culture. The right mix is necessary if we are to be "fully human." He suggests the first priority is protection of wilderness at every scale from the continental to the vest-pocket backyard. Wild nature is the life source and those postmodernists who deny this reality are simply wrong. Earth's Edens are fast becoming isolated islands in a sea of human activity, so hedge them against further exploitation. Do not waste time trying to protect endangered species but go for the preservation of whole landscape ecosystems. National Parks are not enough. These "wildernesses in boxes" must be connected by corridors for migration and gene flow (like the recently protected Tatshenshini watershed in northern British Columbia connecting wild lands in Alaska with those in the Yukon Territory). At the local level, plant the grassy lawn to wild flowers, and discontinue mowing and spraying. Nurture additional natural parks, hedgerows, greenways, green belts, untrammelled river valleys. "If we can set aside sevenths of our time for holiness—that is, for purposes higher than human aggrandizement—why not sevenths of our spaces?"

Protecting ecosystems is the first priority but, he says, that does not mean neglect of civilization and the city. Though wilderness must be protected from the city, the latter needs no protection from wildness that should be encouraged to send its green tendrils into the urban environment. This said, he warns against the concept of "organic cities," because cities are not meant to be "organic." They are inorganic creations where steel and stone, bricks and glass and asphalt constitute the matrix. Humans are preadapted to the urban life, taking to its abiotic technology, its machines, its violent traffic, "its thundering herds of buses, taxis, and humans," like foraging ancestors took to the thundering herds of the Serengeti.

Eisenberg visualizes the city as a place sealed off from the messiness of nature where the human imagination "can play with blocks—with hard man-made objects . . . where technology is given free reign to make life as convenient, fast-paced, baffling, luxurious, tough, or decadent as it likes. . . . It is not that interesting tidbits of human culture cannot be gleaned in the suburbs or in rural villages; it is just that the city gives you more bang for the block."

Therefore, the urban milieu is not a place from which you must escape in order to live a "fully human" life. Cities spring from our nature and so are natural. Cities can meet our quirky needs, he says, and then, optimistically, "without doing too much harm to nature." In fact, well-planned cities can protect nature. The City Tower can be the Eden Mountain's best friend—if it concentrates humans and their cultural energies in a small, bounded, insulated place, "so that wild nature need not take the heat." Clearly Eisenberg has not assessed the gargantuan appetites of cities on the make, the "vibrant cities" extolled by Jane Jacobs.

Many patterns of human living show that people do not like large doses of wild nature, and the same goes for too much civilization. Consciously or unconsciously we try to balance the poles of our nature, seeking a marriage of the two. In Eisenberg's words we fantasize about having a house whose front door opens on Broadway and whose backdoor opens on the Canadian Rocky Mountains. Well-off people have always been able to enjoy the best of both worlds and in asking the Kantian question—"If everyone did it, would it be disastrous?"—he reveals, if not the answer, at least the troubled conscience of one of the fortunates.

Barry Lopez, in *Crossing Open Ground*, wrote that one of humanity's great dreams "must be to find some place between the extremes of nature and civilization where it is possible to live without regret." That geographic half-way home is Arcadia, the "middle landscape," place of the pastoral life. Humans have attempted various mixes of the wild and the civilized in their search for Arcadia, for Eden partly tamed, where Pan—half goat and half human, half wild and half civilized—plays breezy melodies on his reed pipes. Arcadia can be a new land, a frontier, partly wild farmland, a garden city, a monastic cloister or walled garden, the bohemian quarter of the city, a forest village, the summer cottage, the border between tilled land and wilderness, the hills where the herdsmen summer their flocks, a romantic land where people play the parts of shepherd, shepherdess, cowboy, cowgirl. It can be a dream-world of another time and place, a Golden Age of contented easy living before the forests were logged.

Alas, no Arcadia survives for long in its pristine state. In all its space-time forms, the ideal of peace and contentment in green surroundings is transient. Inevitably the prolific, technology-loving, self-centred species transforms its habitat, with or without regret, into the less ideal.

Suburbia is today's outstanding attempt to create Arcadia. Initially it is the perfect "edge" habitat, the mean between city and country, preferably with trees and a babbling brook, where humans feel snug in semi-wild surroundings. But cities expand, suburbia spreads, countryside retreats. A "varicosity of roads" spring up to speed car transportation between suburbs and city, the wave of Arcadia generated by the crowded city swells and spreads, and soon wildness is erased on the far side while cities lose their tax base and vitality on the near side. "In suburbia, idyll quickly shades into elegy." When both nature and city retreat, enfeebled, the only adventure left is consumption, at which suburbanites excel.

Let us not be pessimists, says Eisenberg. Arcadia, the "middle landscape," the foothills of Eden, is the valuable place and way-of-living that humanity must ever seek. Recognize that the balance between Mountain and Tower is a moving target as nature and culture evolve in their intertwined ways. Confine cities so that they grow vertically rather than horizontally, apply the "sabbatical" idea of one seventh of every region dedicated to inviolable wilderness, and make the Arcadian transitional zone a working landscape of permaculture (such as Wes Jackson's "edible perennial prairie"), of community-supported agriculture, solar greenhouses, organic food production, small town revival, cluster housing, harvest festivals.

Music is the appropriate parable, where concord growing out of discord, needing new discord to keep it fresh. Like musicians improvising on a tune, building on each other's harmonies, our relationship to the wild and the civilized should be a kind of extemporizing Earth jazz whose silences between the notes are *tsimtsums*, vital do-nothing pauses to give the rest of creation room to breathe. And dancing to the music back and forth between city and country, between town house and summer cottage, may for the blessed be the ideal Arcadian way. An interview with Eisenberg after the publication of his book revealed him returning to New York after a spell in the Berkshire foothills.

Commentary

As an exceptionally thought-provoking book *The Ecology of Eden* invites a critical evaluation of its foundations. Few will argue with what floats on the surface—Eisenberg's statement of the dilemma posed by human needs for both the wild and the tame. His prescriptions in terms of land-use planning make good sense: confine the spread of cities, embed civilization in a global wilderness matrix, plan the in-between Arcadias as aesthetically pleasing, sustainable food-producing areas. All this is ideal, but what could possibly make it happen? Ah, there's the rub. The book lacks depth. A subtle but vital ingredient is missing.

The author pins his faith on sweet reason, assuming that without much change in attitude we can think our way out of current problems. He does not challenge traditional values and apparently assumes that, on their foundation, we can solve present and future environmental crises. We do not necessarily have to live poorer, he says, just smarter. Thus the book implies that appropriate political and economic action will result once humanity's needs for both wild nature and tame culture are widely recognized and accepted.

Is there a misconception here that society's main motivating force is reason? What about the subtle promptings of conventional faiths, the legacy of homocentric philosophers and theologians? They are powerful determinants of how, if at all, we will balance the wild and the civilized.

Eisenberg does not dig down to expose these underpinnings of thought and action, perhaps because he is too busy defending their source. He is particularly upset by the notion that the old Semitic texts have justified humanity's war on the world.

Ever since Lynn White Jr. in 1967 attributed "The Historical Roots of Our Ecological Crisis" to the Judaeo-Christian tradition, theologians have been frantically searching for ecological needles in scriptural haystacks. In joining the searchers, Eisenberg turns up interesting but irrelevant facts. For example, a good deal of animism persists in the Bible; the biblical Hebrews were not nature-haters and they had, in von Humboldt's words, a "profound sentiment of love for nature." Hebrew monotheism inspired a faith in nature's harmony, and although God is not *in* nature He is to nature as Shakespeare is to *The Tempest* (the designer-from-the-design theorem).

Such odds and ends are beside the point. The virtues of religions do not compensate for their deficiencies any more than a criminal wins pardon on the evidence that he has always been kind to his mother. In one way or another all faiths, with the possible exception of the animism of ancestral gatherers-and-hunters, have sponsored a war on the Ecosphere. None of the "world religions" challenge agriculture: the earliest and still the major battle field where ecodiversity and biodiversity are the adversaries. None challenges human population over-growth, a fundamental sin against the rest of creation. Many in addition to Eisenberg, casting about for arguments that might take the heat off theologies in general, have confuted White's thesis with the happy thought that Sumerians, Babylonians, Greeks, Romans, Buddhists, Hindus, Taoists, as well as Aboriginal Peoples, have mistreated their wild lands (along with wild animals and/or their domestic animals) as badly as adherents of the revealed religions. If everybody has done it, from the beginning of recorded history, surely religious faiths are not to blame for ecological misconduct.

In all this scrambling to justify the unjustifiable the point has been overlooked that religions—ancient and modern, Western, Eastern, and in-between—are necessarily unecologic. The world religions are in-turned, and that fact controls their form, function, and inability to change. Their primary concerns are with man (in the gendered sense) and, by paternal extension, with God (or gods), usually involving some form of covenant or salvational bargain. Their chief value, of undoubted merit and importance for our species, is teaching the social virtues of benevolence, justice and mercy. Their scriptures are not ecological texts, nor were they ever meant to be.

Were religions ecologic at their roots, they would be focused outward on the Ecosphere: the surrounding, sustaining reality from which humans evolved along with twenty-five million other organic species. They would have a penetrating vision of this nurturing Earth as senior partner, with people as junior partner. They would take seriously the importance of survival in the here-and-now through a regimen of plain living, rather than carelessly eating up the Earth while obsessing about such fantasies as original sin, reincarnation, and soul-salvation in a fancied hereafter. Eden-wildernesses would be the recognized creative centres, and the humanized surroundings would daily pay tribute to them.

Eisenberg is homocentric. His fundamental leanings can be gleaned from many passages of which I select two. It is a shade ironic, he says, that *neopagans* have taken up the Gaia hypothesis (Earth as alive) for it restores the faith in nature's harmony that moved the great *Christian* ecologists—von Humboldt, Gilbert White, Louis Agassiz, George Perkins Marsh. Indeed, he continues, in its pop form the Gaian faith is as monotheistic as Moses' (faith). "It merely shifts metaphysical gears (from transcendence to immanence) and changes gender." The adverb "merely" is not the appropriate descriptor for recalling divinity to Earth and rescuing it from patriarchy!

A second example is approval of Hasidic master Simha Bunin of Pzhysha's advice to carry two slips of paper, one in each pocket, reading: *For me the world was made,* and *I am but dust and ashes.* The second slip is demonstrably true but the first, for any student of cosmology, geology, evolutionary biology and ecology, is nonsense. Yet it expresses the common view of humanity's extraordinary importance in the eyes of God. The same idea is repeated in the 1994 apostolic letter of Pope John Paul II as he explained the significance of the year of Jubilee: "If in his providence God had given the Earth to humanity, that meant that he had given it to everyone; therefore the riches of creation were to be considered as a common good of the whole of humanity." Among infantile beliefs, the idea that Earth was made for the pleasure and profit of the human species ranks first.

The message is selfish. Eden is to be preserved because of its payoff in clean air, clean water, and gene pools, for the one species deemed to be favoured by God. In Eisenberg's opinion, stated several times in the text, Nature offers humanity no ends but only means to human ends.

Therefore, other-than-human nature will be cared for not because it is inherently valuable, offering itself as a worthy colleague for human collaboration, but because such care is necessary to keep ourselves and our cities alive and healthy. No imaginative glimmering appears in this book as to what Earth might produce in the way of sensitive, emotive intelligence if humans and post-humans decided to partner with Earth and let it lead, say for the next five million years—a mere fifth of one per cent of its remaining life span.

The fundamental cause of global ecological disaster is species selfishness, inturned homocentrism. The best environmental ethic that the homocentric faiths have been able to muster is *stewardship* as defined in the Genesis text: "And the Lord God took the man, and put him into the garden of Eden to dress it and to keep it." Stewardship puts humans firmly in charge as dressers and keepers, as managers tending Earth's assets so as to keep the goods flowing, perhaps putting a bit aside now and then but only because it may be useful some rainy day. Eisenberg is a steward in the planet managers camp, not overly concerned with problems of human over-population, economic growth, and mega-city expansion.

Philosophically, stewardship fails as the guardian of Eden and Arcadia. Humans fixated on their own welfare, whether on their body/minds in this world or on their souls in the next, cannot help but be injurious to the rest of creation. History shows that people have again and again ruined that land that supported them because their leaders lacked foresight or because hard times forced liquidation of "resources" for the gratification of the moment. Today Earth's capital of soil, air, water, plants and animals is sacrificed to make jobs, to provide the necessities for a population out of control, to make a minority rich, to satisfy frivolous wants. The message is writ large in the current economic system that discounts the future and encourages grabbing what you can now, because "for me the world was made" and I want my slice of it right now.

Midway through his book Eisenberg muses, "How on earth are the likes of us going to act as poet-legislators for the planet, or as counsellors in the marriage of man and nature? For all our tramping around in the woods, we are ecological celibates."

"Celibates" is the wrong word. "Ignoramuses" is better, and that descriptor goes for all of us. Would it not be wise, then, in books such as this about human ecology, to turn one's back on the "planet managers" and take a less-biased look at the misnamed "planet fetishers?" At least they recognize the dangers of hubris and the reckless speed of civilization. They believe that Earth—the source and support of organisms—has vast untapped stores of wisdom concerning human nature and human ecology. Without some such universal belief, how can humanity be convinced of the necessity of redemptive *tsimtsums*?

References

[1] To avoid accusation of the same crime, I declare myself an ecologist as deep as I can go, not one of the Deep Ecologists who might more accurately have called themselves "Biocentric Philosophers." Study of the sometimes-science "ecology" has convinçed me of the oneness of nature: the unity of body, mind, and Earth. I am not a believer in the Gaia hypothesis as phrased by Lovelock, for Earth is no mere "organism" but much more. Earth is a supra-organic Being that under beneficent sunlight confers life on organisms such as we. I am convinced that, for sensory humans, the tangible Earth is divinity incarnate, and all elsewhere-ideas of "soul," "spirit," and "God" are diversions from this fount of values.

"NEW ECOLOGY?" GIVE ME A BREAK!

A Review of *Discordant Harmonies: A New Ecology for the 21st Century,* by Daniel B. Botkin, (New York: Oxford University Press, 1990).

Most people today recognize the importance of ecology, yet few books on the subject explain that its practitioners, like geographers, are all over the map. That ecology is an outward-looking *viewpoint*, interested primarily in the relationships between *organic things* and their contexts, is clear enough. A source of confusion is the fact that ecology's organic things, its *subjects*, are various. They can be individual organisms (autecology), groups of similar organisms (population ecology), or all organisms occurring together in the same place (community ecology). Further, when any three-dimensional piece of Earth space with its air, water, soil and organic community is studied as an integrated whole, as a sort of mega-terrarium, then a recent but most important fourth subject—ecosystem ecology or ecosystemology—is recognized. Obviously the ideas and hypotheses brought to problem-solving in each of the four subject-fields differ.

When writers of popular books on ecology fail to say in which of the four ecology fields their expertise lies (whether through ignorance or the misplaced belief that their particular focus is more important than the other three) then gullible readers, especially philosophers and historians, are likely to be misled. Such books are apt to be uncritically accepted as the latest word on what ecology is all about. Particularly dangerous are books that claim to be the vanguard of a new Ecology.

Exhibit #1 for the prosecution is Daniel Botkin's *Discordant Harmonies: A New Ecology for the 21st Century*. Because it presumes to speak for all ecology, it is a thoroughly bad and misleading book.

Discordant Harmonies is focused primarily on problems of population ecology, dealing with the so-far-unpredictable ups and downs from year to year of the numbers of moose and salmon, sea otters and tree wood-increments. Since first printing in 1990, it has become something of a bible for "resource managers"—those who labour to sustain desirable numbers of animals for shooting, trees for clear-cutting, and fish for netting. Its thesis is simple and appealing: because we cannot predict the numbers, the world as we find it is unstable, unpredictable, unbalanced, purposeless, and therefore badly in need of human guidance to shape it to the heart's desire. It is up to us to substitute for Nature's "discordant harmonies" such pop melodies as suit the musical moods of today.

The theme is trendy for who, in these worsening times, does not welcome a feel-good message that lets people off the hook? It justifies our past and present creation of local and global environmental problems because, really, they are not half as bad as what Nature senselessly and capriciously does when left to itself. At the same time it justifies the human penchant for tinkering, fixing, patching, supervising and administering not only Earth's forests, grasslands, fisheries and wildlife but also the matrix elements of air, water, and soil.

The text of Botkin's book is also enlightening for those interested in the close interactive ties between the taken-for-granted values of society and those of the science that serves it. As pointed out by a forest industry man before he enthusiastically takes up Botkin's theme: "Whether we are commercial foresters or ardent forest preservationists, we are all driven by values and biases that are a product of our culture. We always see the forest through human eyes, and even at that, we experience human perception filtered through some very strong cultural lenses."[1]

No field of human endeavour is immune to the "values and biases" of its culture, as we are reminded by philosophers, including those with a particular interest in science.[2] Books by scientists, from Einstein to Botkin, unintentionally reveal "human perception filtered through . . . strong cultural lenses." The old idea of science as pure and dispassionate, as disengaged from the subjectivity of culture and in that sense "objective," is *passé*. Most scientists have yet to recognize this truth, because few scientists have time for philosophy.

Scientific Vision in a World Perceived as Discordant

The vision of the world that Botkin presents is filtered through the astigmatic eyes of a scientist who takes it on faith that variability in species numbers from year to year, or from decade to decade, correctly mirrors the realities of the wider world—that unpredictable organic phenomena, such as the yearly ups and downs of salmon and cod stocks, truly indicate the way the whole world wags. Botkin concludes that human control of the planet is necessary to eliminate Nature's fundamental flaw.

Claiming to lead ecology into the new millennium, the book's tone conveys the unmistakable accents of the late nineteenth century. As if transported back one hundred years, one hears the voice of the Victorian Darwinians whose ideal of civilization involved the vigorous conquest of nature by science and technology.[3] Botkin writes,"We have the power to mould nature into what we want it to be . . . Nature in the twenty-first century will be a nature that we make. . . . We do not take an engineering approach to nature, we do not borrow the cleverness and the skills of the engineer, which is what we must do . . . we need to instrument the cockpit of the biosphere." These messages resonate with the turn-of-the-century forestry philosophy of Gifford Pinchot: control, manage, and use resources prudently. After a hundred years of unwise exploitation, "wise use" is still a popular slogan, justifying the human onslaught.

Populations fluctuate, therefore Nature is Unstable

Botkin has spent his scientific career tracking the ups and downs of species numbers while searching for their elusive causes. Like other population ecologists he has discovered that the numbers of elephants, moose, fish, sea otters and trees are inconstant over time. When left to themselves in nature they do not multiply to some fixed equilibrium point. Rather their numbers fluctuate, not regularly but in a random fashion. Who can tell how many snowshoe rabbits or ruffed grouse will appear in the bush next year, or in the next ten years? Populations tend to be unpredictable, their periodic swings corresponding more closely to the numbers thrown up by chance, as in a game of dice, than to any optimum target.

From this inconstancy he argues that Nature is inconstant, unharmonious and discordant, offering humanity no firm goals, no guidelines by which to navigate. People must take over the planet, wisely and prudently choosing the goals to be pursued, marshalling science and technology to the task of global management. Old myths and metaphors that stand in the way of this "factual" view of the world must be discarded.

Then progress can be made, civilization can be advanced, and the world made comfortable and pleasant for us all.

Note that the whole argument is based on the failure of population ecology to explain the fluctuating numbers of organisms in time, which justifies the pronouncement that Nature is a failure. The bizarre thesis is that science's failure is really Nature's fault. A more reasonable thesis is that Nature is doing just fine, though in complex ways that species numbers do not mirror. Might it be that as proxies for Nature populations strike out?

In most of his chapters, Botkin equates the behaviour of species populations with Nature's ways. But sometimes, thinking a little deeper, he asserts that Nature taken in its largest sense is the biosphere, a point that raises important but unexamined questions: Is the biosphere (the organism-filled skin of the Earth or Ecosphere) inconstant in the same way as the populations of species it contains? What is the relationship between populations and the specific places—the geographic ecosystems—that they occupy?

As many ecologists understand it, the Ecosphere (i.e. the biosphere set in the whole-Earth context) is the planetary life-system, the world as a material entity that has evolved for 4.6 billion years or so. It is a structured object—the gaseous air overlying water and solid land, with organisms clustered primarily at the gas-liquid-solid phase boundaries. Much evidence that all parts are interactive and symbiotic is accumulating. The compositions of atmosphere, hydrosphere and lithosphere show unmistakable signs of organic contributions, just as organisms exhibit clear evidence in their bodies of endowments from air, water and soil. Mentally dividing the Ecosphere into volumetric sectors at various scales, such as oceans, continents, regions and landscapes, aids understanding. If these divisions (ecosystems) are functional parts, they must have the same structure-composition as the Ecosphere, i.e. a volumetric matrix of air over water/land in which organisms are enveloped. Thus each ecosystem is a three-dimensional object amenable to study, a life-filled piece of terrain with a particular geographic position. Such "geoecosystems" are the true "units of nature" on the face of the Earth.[4]

Research studies of landscape and waterscape ecosystems, beginning with those of the International Biological Programme of the late 1960s, have shown the difficulty of understanding these dynamic open systems with their interacting organic and inorganic parts. Their study requires interdisciplinary teams whose members work closely together over long periods of time. More popular with ecologists, for a variety of reasons that include the research granting system and pressure to publish papers, are the less complex and easier to isolate organic parts of ecosystems; that is, groups of individuals of the same or of different species, co-occurring as populations and communities, respectively. But assemblages of organisms as they fluctuate in numbers from year to year are not good surrogates for Earth and its ecosystems. Further, populations and communities—taxonomic assemblages of organisms—are questionable "objects" for scientific study.

Erroneous Acceptance of Populations as Nature's Surrogates

In the context of real Earth spaces (i.e. of sectors of the Ecosphere), any population is a group of individuals of a particular kind considered apart from the functional ecosystem to which, in a life-supporting and life-maintaining way, it is adapted. A "pack of wolves" is a dead abstraction without the tract of forest through which it ranges. The phrase "polar bears" is a dead abstraction without sea-ice and seals. Sight that picks out "figures" against a disregarded "background" allows us to make these theoretical separations, ignoring ecological reality. Living organisms cannot be separated from their environmental matrix.

When groups of similar individuals *are* separated in thought from their environmental matrix they become a taxonomic category or class called a population, usually defined as members of a particular species in fairly close contact with one another. A population is an artifact in the sense that no spatially associated cluster of organisms has an existence apart from the air/soil/water/food of the ecosystem (sometimes called "habitat") that supports it. Phrased another way, a population *per se* is not a structural-functional thing.

Botkin rightly criticized the old idea that a large plant community (called a "formation") is some kind of super-organism, stating, "There is no inside and outside of a super-organism . . . if there are not sharp boundaries then the super-organism cannot really exist." In this passage, "sharp boundaries" are irrelevant (no one doubts the science of soils even though soils have no sharp boundaries), but the inside-outside idea is pertinent. He fails to see the parallel with his studied populations (of spruce trees, rabbits or codfish) which, exactly like the plant community, have no inside or outside. Neither population nor community is an articulated entity. Neither possesses an internal physiology nor an external ecology in the sense of that possessed by a structurally articulated organism, or a three-dimensional, volumetric geoecosystem.

The concept of "population" is useful for such studies as evolutionary genetics, where the focus is on how a species evolves over time. But the concept misleads when accepted as a proxy for the natural world. Uncritical acceptance of populations as fit objects of study rests on the fact that their members can be counted and graphed. Although populations are not *structural/functional* objects they do have *compositional* characteristics that are easily quantified and mathematized. Once they are defined as "lion prides" or "aspen stands," the changing numbers of individuals within the group can be tallied and analyzed for patterns and trends, for constancy and inconstancy, and for correlation with selected environmental "factors." For a hundred years this has been an ineffective academic exercise, a glass-bead game still played in countless university biology departments (for a telling example and a summing up of fruitless "population ecology" see Krebs et al.).[5]

Botkin would improve the relevance of population studies by better techniques of analysis, using probability models and "smart" computers. These too will fail, because inadequate tools or faulty techniques are not the problem. The fundamental difficulty is that the subject is misconceived. No studies of populations *per se* can enlarge understanding of organic processes in the Ecosphere unless accompanied by close attention to the "units of nature on the face of the Earth" that support and sustain them. Only a focus on the dynamic geoecosystems within which organisms have evolved and are sustained will lead to better understanding of populations, communities, and their fluctuating numbers.

Ecosystem Devalued to a Concept, an Idea

One section of *Discordant Harmonies*, dealing with endangered species, inadvertently counters the book's main thesis that Nature's ways are accurately reflected by swings in population numbers. Comparing the imminent extinction of the California condor with the recovery of the whooping crane, Botkin mentions the importance of "habitat." That of the whooping crane is "intact and self-sustaining" in northern Alberta and along the Texas coast while the condor's habitat in California is virtually destroyed. We learn, he says, that "*the condition of the habitat is more important than simple population numbers.*" He continues, "Conservation of endangered species is . . . understood to depend on *the idea of an ecosystem* rather than on simple analyses of populations" (emphases added).

Botkin has it exactly backward. The key to conservation of species in their fluctuating numbers is security of Earth's geoecosystems that house them. Ecosystems are real things, not just ideas, and populations are the abstracted parts. The whooping crane's summer ecosystem is not a vague "idea" of habitat complexity—it is the dependable, constant, calcareous wetlands in Wood Buffalo National Park offering food and nesting sites. For Botkin the complex ecosystem that exists in the north woods—intact, dependable, and self-sustaining—merits no standing in the debate as to nature's reliability, while the crane population—its migratory numbers fluctuating from year to year as it runs the gauntlet of guns and overhead wires—proves again Nature's inconstancy.

Reliance on population numbers, as counted over time to prove or disprove Nature's capriciousness, is unjustified. Populations are not proxies for nature. Botkin's so-called "New Ecology" founders for lack of a firm conceptual foundation.

Nature's More Constant Face: the Importance of Scale

Nature conceived as the Ecosphere and its sectoral ecosystems presents a different face than the populations of organisms assumed to be its surrogates. Consider any regional forestland ecosystem based on particular geological strata, on landforms of long-term permanency whose surface layers are stable soils—a relatively slow-changing stage on which a suitable suite of organisms, fluctuating in numbers, come and go. The above-surface part of such a forestland ecosystem is changeable in terms of seasonal weather (short term), but the climate (long term) shows many regularities as to length of growing season, month of maximum precipitation, annual ratio of precipitation to evapotranspiration, depth of ground freezing, etc. True the floristic and faunistic composition of any patch varies yearly in a minor way during cycles of drought and moistness, and varies by decades in a major way as disturbances such as fire initiate successional stages. Nevertheless the regional forest ecosystem remains a recognizable and somewhat predictable unit balanced between the fixed and the fluid. The fact that from a short-term viewpoint we cannot exactly predict a year or two ahead what the summer rains will be, or how much new wood a forest will produce, does not condemn Nature as inconstant and badly in need of human guidance. The long-term viewpoint reveals at least a relative, slowly changing stability.

The straw-man that Botkin knocks down is the idea of *exact, precise, fixed* states of equilibrium in nature, a viewpoint held by few today. At the same time he admits the lack of clear meanings for "constancy" and "stability" as applied to populations, communities and ecosystems (undefined), "and the difficulty," he says, "increases in that order." Questions of scale both in space and time, important but only briefly mentioned, could have clarified these concepts. For example, any small-scale patch of forest may run through a cycle of succession that makes it seem inconstant, but the large-scale forest region of which it is one mosaic part may exhibit remarkable overall constancy. On a time-scale, weather is inconstant but climate relatively constant. Were Botkin a landscape ecologist or a marine ecologist, rather than a population ecologist, he would not have discovered the degree of capriciousness in Nature that the sole focus on populations reveals.

Managing the World

The cultural milieu from which Botkin's deep beliefs and myths are inseparable is the affluent consumer society of North America. As a card-carrying member he has to reject the idea of an ordered and well-balanced universe because *if there is order in Nature* then humanity's intelligent role is to back off, try to "go with the flow" of that order, interfere as little as possible in the Ecosphere—at least until we understand more about how it functions. Such a radical idea is anathema to current culture. Adoption of it would cause widespread unemployment among scientists, resource managers, technocrats, and willing workers. The fear of job loss perpetuates many kinds of destructive work.

The idea of an inconstant and orderless nature where chance plays the major role fits perfectly with the managerial society whose purpose is to control, exploit and grow. For example, at the moment, forestland ecosystems are poorly understood, yet ignorance is not considered a valid reason for restraint in their use. One might suppose that everything possible would be done in the way of careful forestry to protect forest-ecosystem compositions and structures. Instead, optimistic banners such as "adaptive management" are hoisted as clear-cutting and forestland clearing continue.

A changeable Nature that exhibits no perfectly stable characteristics, no fixed equilibrium points, *demands* intervention everywhere. Therefore Botkin's repetitive message: embrace technology and prepare to manage the world, species by species. "Under the new management," he says, "one starts with the question: how many sea otters are enough?"

Nowhere in the book is the more fundamental question asked: "How many people are enough?" Nowhere does he entertain the idea that humanity's role might be to keep ecosystems healthy, ministering like sensitive gardeners to those we use so that their diversity, productivity and evolutionary creativity—from sea otters to cedar/hemlock forests—can continue in the ancient proven way.

Co-existence with the Ecosphere

Botkin is honest in admitting ignorance of how the "dense complexity" of crowds of species across Earth's surface persist and have persisted for so long. Why the tremendous biodiversity of this marvellous world? One main reason did not occur to him: that through the entire multi-million-year history of the planet no one species has ever before tried to manage all the rest.

Suppose that, twenty million years ago, there had been an intelligent animal species "managing" forest wildlife according to current hit-and-miss standards, while logging the tropical rain forests in which *our* ancestors were evolving? That would have meant curtains for us. In the same way, we are condemning thousands if not millions of other creatures whose potential, if allowed to evolve and develop, might in some future age exceed our own.

The thought is amusing that any particular anatomical part of our body might dictate to the whole, decreeing the fates of other organs in the interests of its own growth. Yet *Homo sapiens*, born from the living-system Ecosphere and simply one of its many organic parts, is seriously setting itself up to manage the whole Planet for its own comfort and convenience, justifying its foolish actions by the myth of an "inconstant Nature." And the only excuse, in this particular book, is the failed-science fact that "population ecologists" cannot predict the numbers from year to year of the organisms they study!

The final sentence in Botkin's book, shorn of the first line, reads like a saving after-thought:

> If nature in the twenty-first century will be a nature that we make, then the guide to action is our knowledge of living systems and our willingness to observe them for what they are, our commitment to conserve natural areas, to recognize the limits of our actions, and to understand the roles of metaphor and myths in our perceptions of our surroundings.

Guiding our actions by knowledge of "living systems," rather than populations, could have been the theme for a truly new ecology.

References

[1] Weyerhaeuser Jr., George H., "The challenge of adaptive forest management: Aren't people part of the ecosystem too?" *Forestry Chronicle* No. 74 (1998) 865-870.

[2] Proctor, Robert N., *Value-free Science? Purity and Power in Modern Knowledge.* (Cambridge, Mass. & London, England: Harvard University Press, 1991).

[3] Worster, Donald, *Nature's Economy, the Roots of Ecology* (San Francisco: Sierra Club Books, 1977). See pages 170-178 for a close match between the Victorians' and Botkin's views.

[4] Rowe, J.S. and B.V. Barnes, "Geo-ecosystems and Bio-ecosystems" *Bull. Ecol. Soc. Amer.* Vol. 75 No.1 (1994) 40-41.

[5] Krebs, C.J., R. Boonstra, S. Boutin, and A.R.E. Sinclair, "What Drives the 10-year Cycle of Snowshoe Hares?" *BioScience* Vol. 51 No. 1 (2001) 25-35.

WHAT HOPES FOR ECOLOGIZING RELIGIONS?

A review of *Spirit and Nature: Why the Environment is a Religious Issue*, Steven C. Rockefeller and John C. Elder, eds. (Boston: Beacon Press, 1992). Reprinted with permission from *The Structurist*, No. 39/40, 1999–2000. University of Saskatchewan, Saskatoon, pp 84–89. Edited for this collection.

Today's traditional religions and philosophies are troubled by Earth's deteriorating environment. The various faiths ask themselves, "How can this new vexing set of problems, obviously important, be addressed within our established, traditional wisdom?" The thoughts parallel those of educators who, after brief thought, add to the school or college curriculum a secondary or tertiary subject named Environmental Studies. Rarely are the taken-for-granted tenets of established educational and religious faith-systems questioned as possible mischievous contributors to the current predicament.

In *Spirit and Nature: Why the Environment Is a Religious Issue*, a group of theologians and philosophers examined the major religions, searching for latent ecological wisdom. The ten chapters present a variety of viewpoints: Native American, Jewish, Christian, Islamic, Buddhist and Humanistic (liberal democratic). From the collection, I have chosen the seven that provide an index of ecological literacy for this important branch of the humanities. Early in the book a warning should have appeared: always expect exponents to put a fair face on their faiths, never to admit deficiencies in ecological understanding. Expect, too, expression of the competitive spirit, especially by triumphalists—those who hold that their faith alone has caught truth by the tail.

Audrey Shenandoah of the Iroquois culture outlines a tradition of thanksgiving to the Creator. She begins by acknowledging the various sustaining elements—mother Earth, elder brother Sun, grandmother

Moon, water, plants, animals, thunderclouds, stars—and emphasizes the importance of language that makes these things easily understandable in the land-based native tongue compared to the more abstract "cold words" of the English vernacular. Change in the individual, she says, has to come from a feeling for the Earth.

These admirable thoughts, inherited from a mainly foraging society that lived close to Earth, are difficult for agricultural/industrial societies such as ours to assimilate. We cannot, I believe, simply adopt traditional ecological knowledge (TEK), no matter what its face value. Only a rationale rooted in our own Western tradition can provide the needed bridge to a closer liaison with the Earth

Ismar Schorsch unintentionally puts his finger on a root source of environmental troubles, describing the foundational Jewish faith as a triumph of spirituality over the senses, as a system of instinct-renunciation, a strain of asceticism blended with a love of learning, a practice of self-restraint that intuitively respects the value and integrity of its natural environment, a regimen of plain living and scriptural study. The Hebrew Bible prompts compassion for animals, he says, and "ideally, Scripture would have us live as vegetarians." Jewish law is rife with regulations designed to rein in "man's unlimited use of his environment."

The two parts of the above message do not jibe. Treating kindly Earth's ecosystems and their constituent organic and inorganic parts will be difficult as long as spirituality is elevated over the senses, mind over matter, and obedience to the scriptural word above all. Much Greek and Jewish thought has pointed people away from Earth.

Sallie McFague reflects on the convergence of ecological and feminist modes of thought, urging a reconstruction of the way Christians understand God, humanity and nature. Her theology shifts the emphasis from authority and hierarchy to participation and community. God is immanent, not external, she says, and we are co-creators in a world whose reality is the contemporary scientific picture. Failure to see our ecological place is the cause of the planet's woes. The central calling of all religions should be a unifying "planetary agenda" that stresses peace, justice and the integrity of creation.

Obviously McFague foresees Christianity oriented in a new direction, similar to the vision of the Roman Catholic priest, Thomas Berry. The sticking point for Christianity, as for almost all religions, is the content of the "planetary agenda." Will it rewrite and reverse the humanity-before-Earth banner under which we now march backward?

J. Ronald Engel's secular faith is spiritual democracy and a commitment to "natural religion" centred on the principles of freedom, equality and community. Nature and humanity will be liberated together by a more profound understanding of citizenship. Drawing his texts from Henry Thoreau and Vaclav Havel, Engel argues that the ecological movement in its most universal form is "responsibility to and for the whole." To doubt that Spirit infuses Nature is to doubt that true citizenship is possible. A second secular voice, Robert Prescott-Allen, advocates good works, especially at the international level, along with acceptance of a world ethic of sustainability.

This is humanism at its best, expressing worthy ideals inherited from the Age of Reason. It is the foremost faith of today's secular society. Unfortunately humanism, as an exclusive focus on people, is the fundamental problem.

Seyyed Hossein Nasr advances a strong critique of Western humanism and secular science which have impacted unfavourably on Islam traditions. Only a spiritual rebirth of the individual, involving a reawakening to the Divine Centre, will bring an enduring solution to current problems. The Quran, "the very Word of God," draws no clear line of demarcation between natural and supernatural nor between the world of man and that of nature. As God's vice-regent, humans are custodians of the world and the world's creatures. They must nurture and care for the ambience in which, playing the central role, they are conduits of divine grace.

Islam joins Judaism and Christianity in the belief that "for me the world was made." Tenzin Gyatso, the fourteenth Dalai Lama, presents a Tibetan Buddhist perspective on spirit in nature, counselling love and responsibility for the whole world. The Buddhist worldview—more a philosophy or way of thought than a religion—emphasizes the immanence of the sacred, and the ecological relatedness of all being. The form or appearance of things is a false front, insofar as it seems to represent independence, for the reality is interdependence. The requisite virtues are compassion, patience, love. Religion is a bit of a luxury, Tenzin says; people can get along without it. But no one can get along without affection and gentleness, which should be extended to the whole world.

Philosophical Buddhism, Taoism and to some extent Hinduism are able to evolve as knowledge about the Earth-humanity relationship develops and changes. In contrast, the revealed religions—religions of the Book—under whose aegis the world is presently run, are in harness to the Word, and their vision is blinkered by ancient scripts.

The convergence of humanity's ancient search for a spiritual centre and the contemporary search for ecological stability is reviewed by Steven C. Rockefeller, professor of religion at Middlebury College and the principal organizer of the symposium. His thumbnail history of Western thought is concise and scholarly, tracing the roots of the environmental crisis both to narrow anthropocentrism (homocentrism) that grants moral standing only to humans, and to utilitarianism that sees only instrumental values (how can we use this to our advantage?) in things other than human. He points out that these ideas are grounded in biblical literature, in Greek thought, and in science. Surveying the historical background of world religions he finds glimmerings of ecological insight in all of them but in none a strong ecological focus—an excellent summation of the thoughts in this volume.

From the foregoing sample it is clear that thoughtful people of all faiths have identified the environmental crisis as a cultural and spiritual crisis. All agree that values and attitudes must change. But on the subject of how to spark the necessary revolution, little is offered except exhortations "to shape up and get spiritual." As usual, the onus is on the person to think better and do better, with little attention to the leadership of deep, taken for granted cultural values. Each exponent of a faith has a nostrum, whether a return to the ancient scriptures or a fuller development of moral democracy, but none, in Rockefeller's words, has "a strong ecological focus."

Distracting from a healthy ecological Earth-view is the solid homocentric foundation of all faiths, a fixation on people as the be-all and end-all of creation. To the question, "What should be our attitude to the world?" the answer almost by rote is, "We should *respect* it," meaning that its priority need not be high. How shall we treat it? We will be *good stewards*, meaning we will continue to be the managers and controllers. How will we develop an environmental ethic? By *extending the idea of community* from family and human race to other creatures, water and soil. In such "ethics by extension" humanity remains comfortably at the centre as God's special vice-regent, manager and administrative expert. The revealed religions are loaded with this restrictive baggage along with a freight of patriarchal nonsense. Reinvigorating such fossil theologies will test feminism, Aboriginal wisdom, and ecology to their extremes.

Theologians of every stripe can discover, somewhere in their traditions, plausible correspondences with today's ecological insights. Yet explicit attention to human-Earth issues is modern, expressed clearly for the first time not in religious literature but in the secular literature of writers such as Henry Thoreau, Aldo Leopold, and Rachael Carson. This suggests that patch-up reinterpretations of traditional wisdom will not fully serve the modern need. A radically new vision of the "good" is required, an innovative faith that is religious in the sense that it is communal, springing from the inner nature of us all as Earthlings, a conversion of minds and hearts to the Earth.

THE RESURGENCE OF THE REAL

A Review of *The Resurgence of the Real: Body, Nature and Place in a Hypermodern World* by Charlene Spretnak, (New York: Addison-Wesley Publishing Company Inc., 1997). Reprinted with permission from *The Structurist*, No. 37/38, 1997–1998. University of Saskatchewan, Saskatoon, pp 90–97. Edited for this collection.

The AGE OF ECOLOGY began with a bang on 28 January 1969, when an off-shore oil well "blew out" in the Santa Barbara Channel. In full view of coastal California, the ocean and its exquisite shoreline were profaned; seals, birds and fish were killed. Here was proof that humans were fouling their Earthly nest. Earlier rumors to that effect had been circulating in such books as Rachel Carson's *Silent Spring* (1962) and Paul Ehrlich's *The Population Bomb* (1968), but Earth Day 1970—triggered by the oil spill—marked the coalescence of a broader spectrum of ecological concerns whose seriousness demanded political action *immediately.* Well, almost.

Several decades later, crises have deepened and forthright political action is still pending. Smog is Earth's airy blanket everywhere. Above it industrial chemicals are eating away the ozone layer—protector of organisms against ultraviolet radiation. As the Earth heats up, so does the debate as to the reality of global climatic change. Doubtful targets for a reduction of carbon dioxide emissions are set at comfortingly distant dates. Nation spars with nation as to who shall kill whales and take the most fish from the seas. Tropical and temperate forests are slashed faster than they are patched. Chemi-agriculture partnered with biotechnology surges ahead, while the quality of food and water declines. Free Trade, the International Monetary Fund and the World Trade Organization combine to weaken environmental regulations worldwide.

The bad news is that those profiting from plunder of the planet mean to keep it that way, even in the face of scientifically verified dangers. The good news is that a more influential chorus of voices now sound the alarms first raised in the '60s and '70s. Back then, the warnings were those of an ineffectual scattering of academic biologists and ecologists. Today, every intellectual worth his/her salt is an ecologist, often with a book to prove it. Psychologists are ecologists, priests and theologians are ecologists, historians are ecologists, physicists are ecologists, feminists are ecologists, poets are ecologists, professors of English are ecologists, ethicists are ecologists, philosophers are ecologists, *even some economists are ecologists*! Will anyone who has something important to say, but doesn't claim to have ecological leanings, please stand up?

Rejoice at this conversion; may it quicken, spread, and bear fruit before the Second Coming. Because EVERYONE *should be* an ECOLOGIST. Human ecology—an understanding of our dependent relationship on Earth—is capable of orienting all human beings to a greater Being than themselves. A rational/emotional "Earth-worldview" is in plain view for everyone who has two feet on the ground and a head in the air (equipped with the standard eyes, ears, and nose), regardless of race, sex or creed. Planet Earth offers new hope for multicultural agreement on a universal ethic.

The above listing of "disciplinary ecologists," wide-ranging though it is, needs further expansion. *All* fields of knowledge should be reworked to reflect humanity's embeddedness in the Ecosphere. Once the critical importance of learning how to live with Earth is recognized, scholarship will be extended far beyond the navel-gazing humanities, to which most of the sciences also belong. Might we not begin to judge the arts, the sciences, and their popular expositions, by the ecological acumen they display? What about an "index of merit" reflecting how well articles or books expose the harmful traditional ideas—philosophic and religious—that have contributed and are contributing to the degradation of Earth and its living contents? A new criticism, ecological criticism, is born!

Charlene Spretnak's *The Resurgence of the Real: Body, Nature and Place in a Hypermodern World*, ranks high on the ecological index of merit. Bypassing idealistic discussions of "reality," she goes straight to the commonsense world. We westerners, she argues, have lost connections to three parts of what is Real: the Body, Nature and Place. By "Body" she means body/mind, a psychosomatic unity that thinks and feels. "Nature" is our total supportive physical setting—planet Earth under sun, moon, and stars. "Place" is the actual part of the Earth each of us inhabits, the bioregion (better "ecoregion") with its particular climate, landforms, soils and organisms. Harmonizing individual and social living with the triple reality of Body, Nature, and Place "grounds" humanity in a meaningful context. After blind centuries when the Real has been ignored and suppressed, she finds optimistic signs of a resurgent interest in Body/Nature/Place—in medicine, in science, in the arts.

Acceptance of the importance of the Real may not come easily, for the dominant Unreal with its supportive centuries-old history still underlies mainstream thought. Spretnak calls the structure of interlocking ideologies that dominates the Western world "Modernity," which allows her to take a few good swipes at today's hyper-moderns who masquerade as post-moderns but are really most-moderns.

Modernity, she writes, preaches both a salvational sense of *progress*, to be achieved by economic expansion and technological innovation, and a dubious faith that these will bring social and cultural advancement in their train. Part of modernity's baggage is denial that the body is important (male philosophers and theologians have elevated mind over body since at least the time of Plato), denial that Nature is important (the Baconian belief that all worthwhile achievements in science and technology are at the expense of Nature which badly needs improvement), and denial that Place is important (cosmopolitan sophistication is valued over the vernacular, with universal standards of culture set in New York, London, and Paris).

Postmodern deconstructionists subscribe to modernity's denial of the Real. Entrenched on university campuses, these most-moderns preach their dogma that such things as Body, Nature, and Place are meaningless "fictive unities," no more than social constructions fabricated by language, serving subtle power plays. Spretnak agrees that individual and collective ways of thinking *are* situated in cultural constructions. But, she says, observe what lies *around* and gives meaning to our artifacts, especially to language. Clearly, cultural constructions are embodied understandings of the Real, identified with a world of Nature, with flesh-and-blood Bodies that exist within bioregional Places. When deconstructionists solemnly announce, "Nature is what we make it," they prove themselves bad ecologists. They do not understand that the supportive and creative *context* of individuals is far more than society, culture, language. Body, Place, and Nature provide the images and metaphors by which we understand each other. The Real is the source of language, rather than language creating the Real.

How did we ever come to perceive body/mind as two separate entities, Nature as a dead resource, and Place as inconsequential? Four intellectual movements, beginning in the fifteenth century, share the blame. The Renaissance made Man the measure of all things and set homocentric humanism on its way. The Reformation formulated a theology of the individual that elevated the male while devaluing both the female and an Earth that was no longer a source of divine revelations. The Scientific Revolution quantified a mechanical world, extending male power over nature while women suffered the sexual terrorism of the sixteenth and seventeenth centuries. The Enlightenment consolidated the secular, individualistic worldview. Modern Man emerged as the detached manipulator of the natural world. And so modernity, ever since the Renaissance, has continued and inflated the bad habits of Western thought that began when Greek rationalism and idealism replaced an organic and holistic view of the world.

Spretnak advocates Ecological post-modernism that posits an unbroken continuity of cosmos, Earth, continent, nation, bioregion, community, neighbourhood, family, and person. These, she says (taking a leaf from the deep ecologists), are the extended boundaries of the self, the most comprehensive sense of our being. "Earth is the great economy, the great educator, the great healer, the great organizer, the great artist and storyteller, the great experimenter, the great blend of cosmic novelty and continuity." Our groundedness, our freedom, is *in* Nature where all are embedded in orderly, bodily, ecological-cosmological processes. Modernity has imposed *its* mechanistic order on these processes, as it has over the centuries on women. Hence to be truly post-Modern is to be ecological and feminist—a calling open to all genders.

Those familiar with the author's previous works may be surprised that in this book "feminism" *per se* is somewhat muted (the term does not even appear in the index). Spretnak has chosen not to rub salt in the wounds of the "revealed" patriarchal religions that she so vigorously excoriated in earlier writings. All the put-downs suffered by women are now identified as a subset of a larger problem: modernity and its rejection of the Real. It seems to me that a parallel advance will take place when "social ecologists" transcend their narrow focus on human hierarchical arrangements. Ecology should be a constant reminder of the big picture, a revelation of Earth as the essential source and context of all organisms. The meanings of "eco," "ecologist" and "ecology" are belittled when appropriated by humanism's various warring sectors and used as brickbats in their squabbles.

Spretnak's book ranges over the entire cultural spectrum in a learned way, elucidating and criticizing the supportive contributions to modernity of science, education, economics, and the arts. Of the latter, showing her art historian side, Spretnak devotes the book's second half to the Romantic reaction against middle-stage modernity during the

Industrial Revolution in Europe (1775–1830), and to subsequent artistic movements that, knowingly or not, were relatives of it. Coleridge, Goethe, Shelley, Byron, Blake, Schubert—these were the *Realists* of earlier times, protesting against the mechanistic worldview, championing ecological-spiritual sensitivities, convinced that the vital powers of poetic-artistic *imagination* could orient humans in restorative directions. Like others who questioned modernity they earned the pejorative title "Romantics"—today's equivalent of dreamers, "greens" and "preservationists."

She also evaluates more recent challenges to the modernity worldview: the Arts and Crafts Movement (1875–1920), with whose outstanding figures—Ruskin and Morris—she is particularly taken, the counter-modern cosmological and spiritual quests in schools of painting (1890–1951), Counter-Modern Modernism (1905–1939), and the latest lamentable failure, the Counterculture (1966–1972), fated to be superficial because ignorant of its historical antecedents. Tucked in with the latter is a short review of India's falling away from Gandhi's Constructive Program (1915–1948) which, had it survived, could have been the model for decentralized political structures. The theme reappears in the last chapter as an "eco-utopian" vision of the near future through which Spretnak leads her reincarnated hero, William Morris. Because most of us are fated to be urbanites, she has created a tastefully designed green city within an equally green ecoregion where Body, Nature, and Place are given their just dues. A favoured educational book in her Utopia is titled (you've guessed it), *Pioneers of Ecological Postmodernism.*

Spretnak has pulled together in an understandable way many of Modernity's social, political, economic, artistic, and scientific thread-bare weavings. Her critical overview is exactly the background needed for judging, in an ecological/contextual way, the works of others who target more limited topics. Here is a solid contribution to the new ecological criticism.

KEN WILBER, ON TRANSCENDING THIS POOR EARTH

A review of *A Brief History of Everything* by Ken Wilber, (Boston: Shambhala, 1996).

"Nietsche could deny any form of transcendence, whether mortal or divine, by saying that transcendence drove one to slander this world and this life. But perhaps there is a living transcendence of which beauty carries the promise, which can make this mortal and limited world preferable and more appealing than any other" (Albert Camus).[1]

Among the chief foes of Earth are those who would transcend the promise of its beauties and find their heaven in realms more spiritual than "this mortal and limited world." Ken Wilber, the Californian guru, is such a one. Why then, in this book of essays on ecology and Earth, pay him any heed? The reason is that he has subverted a good ecological concept, the "holon." I want to rescue the "holon" from the blunder made when its original meaning was extended from the structure of organisms to the structures of everything else. The correction makes nonsense of Wilber's Earth-devaluing philosophy.

Introduction

Favourable and unfavourable references to Wilber's transpersonal psychology and philosophy had flickered across my peripheral vision from time to time, the most scholarly by Gus diZerega[2] on the negative side and by Michael E. Zimmerman[3,4] on the affirmative side. Therefore I gazed with anticipation at the first edition of *A Brief History of Everything* whose cover features an immodestly large close-up of the author: clean shaven right over the top of his pate, bespectacled dark eyes, and a straight-across mouth with just the smidgen of a smile at the corners that gives the tout ensemble a wise-guy look, possibly unintended.

Not far into the book I recognized a Marxian faith in historical determinism, faith in a progressive path onward and upward that doubtless appeals to aging boomers in today's troubled world they have helped to create. The philosophy of perennial progress can comfort both the well-to-do and the poor, for it justifies the *status quo* as a necessary transitional stage toward the better tomorrow. This happy message is delivered, surprisingly, in tones of defensive belligerence not easily reconciled with Wilber's objective of showing, by personal example, the means of attaining higher levels of consciousness. Further, deep ecologists, ecofeminists and other such are lumped in the category "retro-Romantics" and identified as unwitting foes of human advancement.

In the second edition of his book (published in 2000), Wilber's tone is less acerbic though his message is unchanged. The chief accusation he brings against his thinking opponents is their fascination with this world and their shallowness in paying little attention to the obligatory six or seven transformations of the inner self en route to the world soul. Central to his philosophy are these serial transformations—perhaps a useful personal discipline though one that, by its in-turned individualistic emphasis, is inadequate for effecting substantial changes in social and ecological outlook.

The fundamental flaw in Wilber's philosophy emerges in the first few chapters. His entire system is precariously balanced on the "holon" idea, adopted uncritically from Arthur Koestler[5] who blurred the original usefulness of his neologism by including within it all sorts of dissimilar categories.

In this review my aim is to unscramble Koestler's definition of "holon" and Wilber's use of it. The source of the concept can be traced to the organism and its anatomical structure, and at the end of this article I argue that Earth's ecosystems and Earth itself are logically the inclusive entities that stand above the organism-as-holon, thus exceeding the human-holon in importance.

Koestler's Holons and General Systems Theory

The "holon" is a fine idea because it expresses in one word the truth that each material thing is both a whole to its parts and a part of the greater whole that surrounds it. For example, each one of us is a person-holon, both an individual whole consisting of parts (skin, bone, organs) and at the same time a dependent part of an enveloping Earth ecosystem whose matrix sustains us.

Koestler first used the term "holon" in 1967 in his book, *The Ghost in the Machine*. The following year he organised the Alpback Symposium (titled, "Beyond Reductionism—New Perspectives in the Life Sciences") where he presented a paper, "Beyond Atomism and Holism—the Concept of the Holon." As reported in *Janus*[6] his presentation began as follows: "This is going to be an exercise in general systems theory—which seems to be all the more appropriate as Ludwig von Bertalanffy, its founding father, sits next to me. It seems equally appropriate that I should take as my text a sentence from Ludwig's *Problems of Life;* it reads, 'Hierarchical organisation on the one hand, and the characteristics of *open systems* on the other, are fundamental principles of *living nature*'" (emphasis added).

The terms "open systems" and "living nature" indicate that general systems theorising initially took its cue from organisms and organic phenomena. "Open systems" are those that, contrary to closed systems and the Second Law of Thermodynamics, do not "run down" (increase in entropy) because their openness allows outside energy to enter and keep them going. The physicist Schrodinger called them "negentropic."[7] Examples are tornadoes, whirlpools, and organisms. Further, the term "living nature" refers to organisms, to Earth's biota. Thus the structures of organisms, not the dissimilar structures of societies and cultures, provided the pattern or paradigm for general systems theory—a most important point.

In explaining general systems theory, and as a criticism of reductionism in science, von Bertalanffy wrote, "Reality, in the modern conception, appears as a tremendous hierarchical order of organised entities leading in a superposition of many levels, from physical and chemical to biological and sociological systems. Unity of science is granted, not by a utopian reduction of all sciences to physics and chemistry, but by the structural uniformities of the different levels of reality."[8] The second sentence, on the unity of science, makes good sense in arguing that we should study and understand the world as it presents itself to us, not by reducing everything to molecules and atoms. But the first sentence—in its end reference to "sociological systems"—warrants a warning flag. Sociological systems—such as family, tribe, or nation—differ from physical, chemical and biological systems in that they are strictly people-based. They are population categories, taxonomic groupings of similar things, different in empirical content from an organic system such as a plant or an animal that is composed of organs, tissues, cells, and cell parts (organelles).

Koestler pointed out that related holons can be arranged in series of increasing complexity—such as atom, molecule, crystal, rock—calling each such series a "holarchy." A holarchy is a hierarchy of holons, and each holon, he said, looks both up and down the hierarchy. It is two-faced like the Janis-masks of theatre—a whole to its parts below and a part to the whole above. All reality consists of relational holons, not separate "things." This useful concept dissolves the antagonism in science between reductionism and holism, for reduction is a way of understanding that moves downward in holarchies while holism is the upward view. Koestler acknowledged that similar ideas, called "levels of organisation" or "levels of integration," had been proposed much earlier by J.H. Woodger (1929)[9] who drew on the organismic philosophy of Alfred North Whitehead.

Holon Concept Fits Material Things

Koestler used the biological model of an organism to illustrate the holarchy concept. His schematic representation took the form of a pyramid or inverted tree, the broad base comprising many sub-atomic particles, merged at the next higher level into fewer atoms, these into fewer molecules, these into organelles, cells, tissues, organs, organ systems, and at the top of the holarchy, the organism. Each level or holon is a whole to the parts below, and a part of the whole above. For example, an organ such as the heart is both a whole to the lower-level cellular tissue of which it is composed, and a performing part of the higher-level organism. This simple biological "tree diagram"—a useful abstraction of the organism's anatomy composed of parts within parts within parts—was then imprudently extrapolated to include psychological and social/cultural phenomena.

In his words, "All complex structures and processes of a relatively stable character display hierarchic organisation, regardless whether we consider galactic systems, living organisms and their activities, or social organisation . . . The tree diagram with its series of levels can be used to represent the evolutionary branching of species into the 'tree of life'; or the stepwise differentiation of tissues and integration of functions in the development of the embryo." He continued, "Anatomists use the tree diagram to demonstrate the locomotor hierarchy of limbs, joints, individual muscles, and so down to fibres, fibrils and filaments of contractile proteins. Ethologists use it to illustrate the various sub-routines and action patterns involved in such complex instinctive activities as a bird building a nest; but it is also an indispensable tool to the new school of psycholinguistics started by Chomsky." In summary, "Thus organelles and homologous organs are evolutionary holons; morphogenetic fields are ontogenetic holons; the ethologist's 'fixed action-patterns' and the subroutines of acquired skills are behavioural holons; phonemes, morphemes, words, phrases are linguistic holons; individual, families, tribes, nations are social holons" (p.37).

No evidence is presented by Koestler to show that all these different phenomena fall into similar, parallel, logical holarchies. In fact they do not. By ascribing hierarchic organisation to "all complex structures and processes of a relatively stable character," the whole of reality as we know it is assumed to be patterned on the structure of the organism. Not so. The anatomical structures of organisms may provide a useful analogy for thinking about other systems and their structures, but organisms are not homologous with all conceivable systems. Significant differences in content and structure exist between organisms and such other systems as languages, philosophies, cultures, customs, economies, agricultures, climates, and so forth.

For example, holarchies that line up the developmental stages of organisms in time sequences do not express the same rules as organism-based holarchies which, by their organelles, cells, tissues, and organs, show the spatial "boxes within boxes" arrangement. Thus development of the human individual from fertilized egg to adult does not parallel the holarchic anatomy of the adult body with its many feedback mechanisms. The foetus is not first organelles, then cells, then organs, then a baby. Neither are evolutionary holarchies homologous with the structures of organism systems. Koestler accepted these quite different holarchies as congruent, referring without discrimination to the hierarchical organization of structures (spatial objects) and of processes (temporal events). To accept such things as mind/brain evolution, or the successive stages of cultural development over the last fifty thousand years, as parallel to the anatomy of organisms, deprives the terms "holon" and "holarchy" of useful meaning. Further, it posits a fictitious and misleading knowledge of the mental and cultural evolution of *Homo sapiens*.

Illogical Holarchies

The fallacy of mixing different categories, and treating them as of the same form, isomorphic, traps many otherwise clever minds. For example, Maturana and Varela[10] discussed organisms, ecosystems, and societies as "multicellular living systems." They did not recognise the fundamental difference between societies, as populations of species, and organisms which like geographic ecosystems are volumetric things. Not surprisingly the only commonality they found among the three was "operational closure"—a suitably vague phrase. Nevertheless, using the autonomy of components as a guide, they arranged from low to high the confused sequence: organisms, animal societies, ecosystems, and human societies. By this kind of thinking, human societies are more important than the Earth that brought them into being and supports them.

Following the same line of thought, Capra[11] also placed human societies at the highest system level because, he said, *they exist for their components* (individual persons), unlike the components of an organism that exist solely for the organism's functioning. According to Capra, good social systems (human societies) are those that exist for the welfare of their components, and bad social systems are patterned on the organism—which is a fascist model for the fascist state! Part of the problem, elevating human society above all else on Earth, is the faulty concept of "ecosystem" held by Maturana, Varela, and Capra. To them it is a hazy equivalent of the multi-species community, just another kind of society that conceptually provides no assistance in sorting out "multicellular living systems." When "ecosystem" is understood to be a volumetric sector of Earth including the organisms that it envelops and supports, its status rises above that of human societies.[12, 13]

No one "general systems theory" exists today because, since von Bertalanffy's time, too many different kinds of systems have been revealed. Perhaps his emphasis on one inclusive theory, and his superimposing sociological systems on biological systems, are to blame for the ease with which many, including Wilber, have made the illogical leap from organism systems to the unlike systems of society, culture, mind, and even consciousness expansion. The Nobel Laureate, Peter B. Medawar, strongly criticised Koestler for building an illogical hierarchy of holons from non-homogeneous elements. Shaky logic of this kind, he said, can be mischievous.[14] Medawar's apprehension is justified by the uses to which Koestler's concepts have been put by Wilber, for the latter's holarchies are illogical evolutionary time-sequences of consciousness, mindfulness, culture, and social structures, all apparently assumed to be isomorphic with the spatial biological "tree" model.

The Wilberian System

Wilber's philosophy is convoluted and cannot here be explored in detail.[15] The flavor of his philosophy is conveyed in the four paragraphs that follow.

All reality (physical and mental) is an assemblage of whole/parts or "holons," each composed of lower-level holons and themselves parts of higher-level holons. All things and processes, symbols, images and concepts are holons (emphasis added). It's holons all the way down and all the way up, according to Wilber. His particular focus is on human evolutionary holarchies viewed as historical, progressive, and purposeful.

"Spirit" is the underlying reality, seeking to actualize itself in higher and higher forms of consciousness. Evolution is a progressive process whereby more complex holon-levels emerge as manifestations of Spirit. From atoms came molecules, from molecules cells, from cells came organisms, and so on up to the complexity of the human mind/brain and on to the higher, mystical, transpersonal levels of Spirit. These higher levels hint at what collective evolution has in store for humanity's future.

Two spiritual paths are the ascending (transcendent) seeking the one, and the descending (immanent) embracing the many. Non-dual traditions attempt to integrate the one and the many, and this is the ultimate saving Way. Wilber states, "Those who do not contribute to this union destroy the only Earth they have." Modernism has emphasised the descending path, the material path, producing today such imperfections as the Gaia religion whose God is "clunking around in our visual field." Wilber is on the ascending, transcendent spiritual path.

Geological and human history demonstrate aspects of the progressive evolution of holons, from the physical to the biological, from simple organisms to the complex, from the conscious animal to the self-conscious human. The "great chain of being" of Plotinus prefigures the dynamic evolutionary sequence that moves upward from matter (the physiosphere) to living organisms (the biosphere) to human consciousness (the noosphere), then on to Soul and finally to pure Spirit (aka God, the All, Emptiness).

Although he recognizes that there is validity to the descending spiritual path with its care for the immanent world, Wilber is clearly a Platonist whose realities are transcendent and immaterial. His evolutionary sequences are linked to the thesis that holarchies possess a spiritual entelechy or guiding principle of development, that history shows a steady advancement from simplicity to complexity and from matter to mind.

The idea that humanity is on an upward progress path is appealing. The main support for the faith in inevitable progress is palaeontology and the fossil record which, over millions of years, generally yields a story of increasing organic complexity with many stops and starts. Is the human race on this cosmic roll? As far as can be told from skeletal remains, organic complexity in the human race has not changed for thousands of years, and the jury is still out on whether various cultural/social stages in the last fifteen thousand years are "more advanced" than others. Is industrial society an advance on horticultural society, and it on foraging society? Much depends on the standards of judgment, such as the work/leisure time ratio, standard of living, population numbers, life expectancy, drain on and damage to Earth, and so forth. In the last forty years new ecological standards, pertaining to sustainability and preservation of Earth's ecosystems and their biodiversity, have come into focus as most important. They indicate that the ideas of Aboriginal foraging societies are more relevant to sustainable ways of life on Earth than are those of the current informational society.

Wilber lines up evolutionary history as proceeding from matter to body to mind to soul to Spirit. As a variant on the same theme, he identifies the evolutionary sequence as progressing from physiosphere (the material domain) to biosphere (the living organism domain) to noosphere (the domain of conscious mind) and on to the theosphere (spiritual domain). He argues that because each emergent holon transcends and includes its predecessor, the biosphere transcends and includes the physiosphere and is itself transcended and included, as "a lower level of structural organization," in the noosphere. From this it follows that because humans, as conscious beings, are the chief constituent of the noosphere, "the biosphere is literally internal to us, is part of our being." We need, he says, an approach that transcends and includes ecology precisely because the noosphere (mind) transcends and includes the biosphere (mere body).

Such faulty arguments assume that the structural organization of physical, biological, and mental categories is the same, providing the bootstraps by which Wilber lifts all reality into aspects of consciousness on their way to pure Spirit. The grain of truth in his dogma is that all organisms are "open systems," constantly internalising energy and materials from the biosphere, but the conclusion that such common sense phenomena as the physiosphere and biosphere—the Earth realities of air/water/soil landscapes in which humans exist—are interior, structural parts of the mind/noosphere will only convince blind idealists.

Four Dimensions of the Personal Self

Wilber develops a more detailed explication of his philosophy around the central theme of Spirit manifesting itself in four interconnected, developmental holarchies: two mental/interior and two physical/exterior. The "person" or "self" derives its meaning from four realities or dimensions: (1) the "me" consciousness, (2) culture (especially the formative language milieu), (3) the way the physical world is perceived, and (4) the social institutions and built environment that frame daily living. Change any of the four and the configuration of "self" changes. In short, the mind ("I"), culture ("we"), and both ("it") worlds of nature and societal structures participate in "selfhood."

This theme is a valuable contribution, and to Wilber we are indebted for the clarity it brings to classifying the prophets and problem-solvers who attempt to better the human condition. Each of the four dimensions or quadrants of the "self" has its champions, either accenting reform of consciousness (Freud), or culture (Weber), or concepts of nature (Skinner), or social institutions (Marx), rather than recognizing the importance of all four. The popular chorus in today's individualistic society is "Change your personal consciousness!" This egoic focus is counterproductive when it diverts attention from serious cultural and environmental issues. The popular thesis, that only self-improvement (self-realisation, self-development, spiritual growth) will change the world, is one quarter correct.

Having formulated the revealing "four dimensions of self" idea, Wilber proceeds to accent one dimension only. He is essentially a psychologist, as his primary focus on mind, consciousness, soul, and Spirit indicate. Wedded to the inner "mind-consciousness" quadrant and its transpersonal possibilities, he is intolerant of those who believe that an outward focus on the world as sensed may also encourage a "spiritual upward" step. Wilber's answer to the ecological crisis is spiritual development on the one path that he has marked out. He criticizes what he calls retro-romantics, ecofeminists, ecomasculinists, for regressing to the childish, pre-rational levels of culture: the Archaic, Magic, and Mythic. They yearn nostalgically for the foraging or horticultural or early agricultural societies, for a return to nature. But, according to him, "The closer you get to nature, the more egocentric you become" (p.319).

Wilber's faith in the evolution of consciousness toward Spirit, coupled with the dogma that the path of human progress has already been mapped out by ancient mystics, provides his justification for denigrating those exploring other paths for the betterment of human life on Earth. Stressing the mystical consciousness, he depreciates the physical and natural. Viewing all realities through the lens of humanism, he cannot conceive any other source of values. He announces the end of evolution as already achieved in pure spirit by a talented few mystics, ancient and modern—and all this based on a false homology between questionable historical/evolutionary series and the holarchy of holons that complex organisms display in their anatomy.

An Earth-based Philosophy derived from the Organic Holarchy

More realistic than Wilber's by-guess-and-by-God evolutionary scheme is an ecological holarchy in the here and now, based on awareness of the close ties between people and the magnificent and beautiful organic/inorganic Earth, plus whatever spiritual overtones their intuitive and intellectual sensibilities bring to it. To quote Capra[16] (p.7), "When the concept of the human spirit is understood as the mode of consciousness in which the individual feels a sense of belonging, of connectedness, to the cosmos as a whole, it becomes clear that *ecological awareness* is spiritual in its deepest essence"

Wilber acknowledges the existence of an ecological holarchy, though stating it wrongly, "Even the anti-hierarchy ecophilosophers offer their own hierarchy," he writes (p.72), "which is usually something like this: atoms are parts of molecules, which are parts of cells, which are parts of individual organisms, which are parts of families, which are parts of cultures, which are parts of the total biosphere. That is their defining hierarchy, their defining holarchy, and except for some confusion about what "biosphere" means, that is a fairly accurate holarchy."

Nonsense! No "fairly accurate holarchy" would mix categories as Wilber does, jumping from individual organisms to families to cultures to the biosphere. Medawar castigated Koestler for precisely this logical error. The first part of Wilber's statement, up to "individual organisms" is correct, as is his statement that the term "biosphere" is confusing. It should be replaced with the term "Ecosphere" or Earth.

A logical ecological holarchy follows the simple principle of containment; that is, each level in the sequence is enveloped as a physical volumetric part by the next higher level. On the analogy of Chinese boxes that fit within one another, each higher level is the environment of those below. This is the sequence that Koestler accurately showed as the pyramidal or inverted "tree diagram" with the organism at the summit. Let us now take it one step further.

From the base upward, atoms are parts of molecules, which are parts of cells, which are parts of tissue/organs, which are parts of organisms, *which are parts of geographic ecosystems, which are parts of the Ecosphere*. Each higher level is the environment or "field" of the ones below, and each lower level is a functional part of the levels above. Note that in this sequence human organisms appear as one among many species/parts of the sectoral ecosystems that Earth comprises. Humans are made from and sustained by the living planet. Physically and mentally they are Earthlings. Truly they are marvellous creatures, but not the be-all and end-all of creation.

Some may suspect hidden in this argument the dangers that Medawar and Capra foresaw in Koestler's "mischievous" social holarchy where the person-as-holon is inferior to the higher-level state and therefore bound to serve its dictates. They may ask, is not the holarchy that places Earth and its geoecosystems above people just another path to totalitarianism? Does it not set strict limits on personal liberty? The question has purchase in today's individualistic culture where any constraints on human freedom to reshape the world arbitrarily as desired are viewed if not as traitorous at least with alarm. In the humanistic tradition, only people possess high intelligence, are important, and loved by God. To suggest otherwise, to place higher values on Earth's ecosystems than on *Homo sapiens,* smacks of "ecofascism."

The "ecofascist" label is loosely used by many, including Wilber. Fascism is a human social institution, a repressive dictatorship backed by force. The creative Ecosphere is not a human institution, neither made by humans nor guided by humans. Humans as Earthlings *are* subservient to Earth and to its sectoral ecosystems from which they evolved, are formed, sustained, and to which they return. That is ecological reality. But Earth's ecosystems express no dictatorial decrees as to human behaviour. Nature/Earth gives humanity free rein to pursue destructive courses without the swift, cruel and vindictive reprisals that characterize fascist nations and international warfare. Earth generally shows humans the folly of their ways slowly, her responses presented as lessons to be learned. Whether Earth is recognised as humanity's body/mind/spirit source and support, and whether or not people act responsibly on that knowledge, is their choice.

In Conclusion

The potentially useful concept of "general systems theory" has faded rather than strengthened over the last fifty years. The probable reason is that the early promoters of the good idea, von Bertalanffy and Koestler, were too ambitious, too hegemonic. They attempted to bring under one roof all kinds of different categories, treating their diverse structures as isomorphic with those of organisms. The result was nonsensical hierarchies or holarchies composed, as Medawar noted, of "non-homogeneous elements." Wilber's philosophic scheme, extrapolating the holon idea from organic structures to fanciful developmental and evolutionary time sequences, illustrates the fallacy.

The holarchy idea makes sense when applied to the structural anatomy of physical things such as the organism—its original inspiration. Extending the logic of containment, the next higher holon or level of integration above the organism is the geographic ecosystem, itself a part of an ecoregion (sometimes called "bioregion")—the chunk of Earth-space, resembling a giant terrarium, within which humans and other organisms exist. The sum of all ecosystems and ecoregions is the Ecosphere, Earth, source and sustainer of life, recognised as superior to *Homo sapiens* whose role, niche, or purpose is revealed as ministering to the health of the more creative Being that envelops it.

By this logic, a respectable rationale is provided for a shift in cultural perspective away from humanity to the marvellous planet that births and nurtures all life forms, a shift from navel-gazing homocentrism to Earth-venerating ecocentrism. Matched with Earth's beauty, this is a transcendence that Camus, in the introductory statement, would approve. Without such a change in perspective, without this fundamental ecological outlook, humanity's future—already glimpsed as clouded—is certain to be storm-wracked.

References

[1] Camus, Albert, *The Rebel* (New York and Toronto: Alfred A. Knopf, Inc. & Random House, Inc., 1951, transl. 1956) 258.

[2] diZerega, Gus, "A Critique of Ken Wilber's Account of Deep Ecology & Nature Religions." *The Trumpeter* Vol. 13 No.2 (Victoria: LightStar Press, 1996) 52-71.

[3] Zimmerman, Michael, *Contesting Earth's Future: Radical Ecology and Postmodernity* (Berkeley, Los Angeles, London: Universiy of California Press, 1994) 196-217.

[4] Zimmerman, Michael, "Possible Political Problems of Earth-Based Religiosity" *Beneath the Surface: Critical Essays in the Philosophy of Deep Ecology*, eds. Eric Katz, Andrew Light, and David Rothenberg (Cambridge, Mass. & London, England: The MIT Press, 2000) 169-194.

[5] Koestler, Arthur, *Janus, A Summing Up* (New York: Random House, 1978).

[6] Koestler, Ibid.

[7] Erwin Schrodinger, *What is Life? The Physical Aspect of the Living Cell* (London, New York, Toronto: Cambridge University Press, 1944).

[8] von Bertalanffy, Ludwig, "An Outline of General Systems Theory" *British Journal for the Philosophy of Science* No. 1 (1950) 134-165.

[9] Woodger, J.H., *Biological Principles, a Critical Study* (London & New York: Harcourt, Brace & Co., 1929).

[10] Maturana, Humberto, and Francisco Varela, *The Tree of Knowledge* (Boston: Shambhala, 1987).

[11] Capra, Fritjof, *The Web of Life: A New Scientific Understanding of Living Systems* (New York and Toronto: Anchor Books, Doubleday, 1996).

[12] Rowe, J.S., "The Level of Integration Concept and Ecology." *Ecology* Vol. 42 No. 2 (1961) 420-427.

[13] Rowe, J.S., "The Integration of Ecological Studies." *Functional Ecology* No. 6(1992) 115-119.

[14] Medawar, Peter, *The Art of the Soluble.* (London: Methuen & Co. Ltd., 1967).

[15] A longer version of Wilber's critique has been published in the on-line edition of *The Trumpeter* and can be found at http://trumpeter.athabascau.ca/content/v17.1/

[16] Capra, Ibid.

EPILOGUE: A MANIFESTO FOR EARTH

TED MOSQUIN AND STAN ROWE

Why this Manifesto?

This manifesto is Earth-centred. It is precisely ecocentric, meaning home-centred, rather than biocentric meaning organism-centred. Its aim is to extend and deepen people's understanding of the primary life-giving and life-sustaining values of Planet Earth, the Ecosphere. The manifesto consists of six Core Principles that state the rationale, plus five derivative Action Principles outlining humanity's duties to Earth and to the geographic ecosystems Earth comprises. It is offered as a guide to ethical thinking, conduct and social policy.

Over the last century advances have been made in scientific, philosophical and religious attitudes to non-human Nature. We commend the efforts of those whose sensitivity to a deteriorating Earth has turned their vision outward, to recognition of the values of the lands, the oceans, animals, plants and other creatures. And yet, for lack of a common ecocentric philosophy, much of this goodwill has been scattered in a hundred different directions. It has been neutralized and rendered ineffective by the one, deep, taken for granted cultural belief that assigns first value to *Homo sapiens sapiens* and then, sequentially, to other organisms according to their relatedness to the primary one.

The recent insight that Earth, the Ecosphere, is an object of supreme value has emerged from cosmologic studies, the Gaia hypothesis, pictures of Earth from space, and especially ecological understanding. The central ecological reality for organisms—twenty-five million or so species—is that all are Earthlings. None would exist without planet Earth. The mystery and miracle called life is inseparable from Earth's evolutionary history, its composition and processes. Therefore, ethical priority moves beyond humanity to its inclusive Earth home. The manifesto maps what we believe is an essential step toward a sustainable Earth-human relationship.

Preamble

Many artistic and philosophical movements have produced manifestos, proclaiming truths that to their authors were as manifest as their five-fingered hands. This manifesto also states self-evident truths, as obvious to us as the marvellous five-part environment—land, air, water, fire/

sunlight, and organisms—wherein we live, move, and have our being. The manifesto is Earth-centred. It shifts the value-focus from humanity to the enveloping Ecosphere—that web of organic/inorganic/symbiotic structures and processes that constitute Planet Earth.

The Ecosphere is the life-giving matrix that envelops all organisms, intimately intertwined with them in the story of evolution from the beginning of time. Organisms are fashioned from air, water, and sediments, which in turn bear organic imprints. The composition of sea water is maintained by organisms that also stabilize the improbable atmosphere. Plants and animals formed the limestone in mountains whose sediments make our bones. The false divisions we have made between living and non-living, biotic and abiotic, organic and inorganic, have put the evolutionary potential of the Ecosphere at risk.

Humanity's ten-thousand-year-old experiment in mode of living at the expense of Nature, culminating in economic globalization, is failing. A primary reason is that we have placed the importance of our species above all else. We have wrongly considered Earth, its ecosystems, and their myriad organic/inorganic parts as mere provisioners, valued only when they serve our needs and wants. A courageous change in attitudes and activities is urgent. Diagnoses and prescriptions for healing the human-Earth relationship are legion, and here we emphasize the visionary one that seems essential to the success of all others. A new worldview anchored in the planetary Ecosphere points the way.

Statement of Conviction

Everyone searches for meaning in life, for supportive convictions that take various forms. Many look to faiths that ignore or discount the importance of this world, not realizing in any profound sense that we are born from Earth and sustained by it throughout our lives. In today's dominating industrial culture, Earth-as-home is not a self-evident percept. Few pause daily to consider with a sense of wonder the enveloping matrix from which we came and to which, at the end, we all return. Because we are issue of the Earth, the harmonies of its lands, seas, skies and its countless beautiful organisms carry rich meanings barely understood.

We are convinced that until the Ecosphere is recognized as the indispensable *common ground* of all human activities, people will continue to set their immediate interests first. Without an ecocentric perspective that anchors values and purposes in a greater reality than our own species, the resolution of political, economic, and religious conflicts will not be possible. Until the narrow focus on human communities is broadened to include Earth's ecosystems—the local and regional places wherein we dwell—programs for healthy sustainable ways of living will fail.

A trusting attachment to the Ecosphere, an aesthetic empathy with surrounding Nature, a feeling of awe for the miracle of the living Earth and its mysterious harmonies, is humanity's largely unrecognized heritage. Affectionately realized again, our connections with the natural world will begin to fill the gap in lives lived in the industrialized world. Important ecological purposes that civilization and urbanization have obscured will re-emerge. The goal is restoration of Earth's diversity and beauty, with our prodigal species once again a co-operative, responsible, ethical member.

Core Principles

Principle 1: The Ecosphere is the centre of value for humanity

Principle 2: The creativity and productivity of Earth's ecosystems depend on their integrity

Principle 3: The Earth-centred worldview is supported by natural history

Principle 4: Ecocentric ethics are grounded in awareness of our place in Nature

Principle 5: An ecocentric worldview values diversity of ecosystems and cultures

Principle 6: Ecocentric ethics support social justice

Action Principles

Principle 7: Defend and preserve Earth's creative potential

Principle 8: Reduce human population size

Principle 9: Reduce human consumption of Earth parts

Principle 10: Promote ecocentric governance

Principle 11: Spread the message

Core Principles

Principle 1: The Ecosphere is the Centre of Value for Humanity

The Ecosphere, the Earth globe, is the generative source of evolutionary creativity. From the planet's inorganic/organic ecosystems organisms emerged: first bacterial cells and eventually those complex confederations of cells that are human beings. Hence, dynamic ecosystems, intricately expressed in all parts of the Ecosphere, exceed in value and importance the species they contain.

The reality and value of each person's ecological or outer being has attracted scant attention compared to the philosophic thought lavished on humanity's inner being, the latter an individualistic focus that draws attention away from ecological needs and neglects the vital importance of the Ecosphere. Extended to society as concern only for the welfare of people, this homocentrism (anthropocentrism) is a doctrine of species selfishness destructive of the natural world. Biocentrism that extends sympathy and understanding beyond the human race to other organisms marks an ethical advance, but its scope is limited. It fails to appreciate the importance of the total ecological "surround." Without attention to the priority of Earth-as-context, biocentrism easily reverts to a chauvinistic homocentrism, for who among all animals is commonly assumed to be the wisest and best? Ecocentrism, emphasizing the Ecosphere as the primary life-giving system rather than merely life's support, provides the standard to which humanity must appeal for future guidance.

We humans are conscious expressions of the Ecosphere's generative forces, our individual "aliveness" experienced as inseparable from sun-warmed air, water, land, and the food that other organisms provide. Like all other vital beings born from Earth, we have been "tuned" through long evolution to its resonances, its rhythmic cycles, its seasons. Language, thought, intuitions—all are drawn directly or metaphorically from the fact of our physical being on Earth. Beyond conscious experience, every person embodies an intelligence, an innate wisdom of the body that, without conscious thought, suits it to participate as a symbiotic part of terrestrial ecosystems. Comprehension of the ecological reality that people are Earthlings shifts the centre of values away from the homocentric to the ecocentric, from *Homo sapiens* to Planet Earth.

Principle 2: The Creativity and Productivity of Earth's Ecosystems Depends on their Integrity

"Integrity" refers to wholeness, to completeness, to the ability to function fully. The standard is Nature's sun-energized ecosystems in their undamaged state; for example a productive tract of the continental sea-shelf or a temperate rain forest in pre-settlement days when humans were primarily foragers. Although such times are beyond recall, their ecosystems (as much as we can know them) still provide the only known blueprints for sustainability in agriculture, forestry, and fisheries. Current failings in all three of these industrialized enterprises show the effects of deteriorating integrity; namely, loss of productivity and aesthetic appeal in parallel with the continuing disruption of vital ecosystem functions.

The evolutionary creativity and continued productivity of Earth and its regional ecosystems require the continuance of their key structures and ecological processes. This internal integrity depends on the preservation of communities with their countless forms of evolved co-operation and interdependence. Integrity depends on intricate food chains and energy flows, on uneroded soils and the cycling of essential materials such as nitrogen, potassium, phosphorus. Further, the natural compositions of air, sediments, and water have been integral to Nature's healthy processes and functions. Pollution of these three, along with exploitive extraction of inorganic and organic constituents, weakens ecosystem integrity and the norms of the Ecosphere, the fount of evolving life.

Principle 3: The Earth-centred Worldview is Supported by Natural History

Natural history is the story of Earth unfolding. Cosmologists and geologists tell of Earth's beginnings more than four billion years ago, the appearance of small sea creatures in early sediments, the emergence of terrestrial animals from the sea, the age of dinosaurs, the evolution with mutual influences of insects, flowering plants, and mammals from which, in recent geological time, came the primates and humankind. We share genetic material and a common ancestry with all the other creatures that participate in Earth's ecosystems. Such compelling narratives place humanity in context. Stories of Earth's unfolding over the eons trace our co-evolution with myriad companion organisms through compliance, and not solely through competitiveness. The facts of organic coexistence reveal the important roles of mutualism, co-operation, and symbiosis within Earth's grand symphony.

Cultural myths and stories that shape our attitudes and values tell where we came from, who we are, and where in the future we are going. These stories have been unrealistically homocentric and/or other-worldly. In contrast, the evidence-based, outward-looking narrative of humanity's natural history—made from stardust, gifted with vitality and sustained by the Ecosphere's natural processes—is not only believable but also more marvellous than traditional human-centred myths. By showing humanity in context, as one organic component of the planetary globe, ecocentric narratives also reveal a functional purpose and an ethical goal, namely, the human part serving the greater Earth whole.

Principle 4: Ecocentric Ethics are Grounded in Awareness of our Place in Nature

Ethics concerns those unselfish attitudes and actions that flow from deep values; that is, from the sense of what is fundamentally important. A profound appreciation of Earth prompts ethical behavior toward it. Veneration of Earth comes easily with out-of-doors childhood experiences and in adulthood is fostered by living-in-place so that landforms and waterforms, plants and animals, become familiar as neighbourly acquaintances. The ecological worldview and ethic that finds prime values in the Ecosphere draws its strength from exposure to the natural and semi-natural world, the rural rather than the urban milieu. Consciousness of one's status in this world prompts wonder, awe, and a resolve to restore, conserve, and protect the Ecosphere's ancient beauties and natural ways that for eons have stood the test of time.

Planet Earth (and its varied ecosystems with their matrix elements—air, land, water, and organic things) surrounds and nourishes each person and each community, cyclically giving life and taking back the gift. An awareness of self as an ecological being, fed by water and other organisms, and as a deep-air animal living at the productive, sun-warmed interface where atmosphere meets land, brings a sense of connectedness and reverence for the abundance and vitality of sustaining Nature.

Principle 5: An Ecocentric Worldview Values Diversity of Ecosystems and Cultures

A major revelation of the Earth-centred perspective is the amazing variety and richness of ecosystems and their organic/inorganic parts. The Earth's surface presents an aesthetically appealing diversity of arctic, temperate and tropical ecosystems. Within this global mosaic the many different varieties of plants, animals, and humans are dependent on their accompanying medley of landforms, soils, waters and local climates. Thus, biodiversity, the diversity of organisms, depends on maintenance of ecodiversity, the diversity of ecosystems. Cultural diversity—a form of biodiversity—is the historical result of humans fitting their activities, thoughts and language to specific geographic ecosystems. Therefore, whatever degrades and destroys ecosystems is both a biological and a cultural danger and disgrace. An ecocentric worldview values Earth's diversity in all its forms, the non-human as well as the human.

Each human culture of the past developed a unique language rooted aesthetically and ethically in the sights, sounds, scents, tastes, and feelings of the particular part of Earth that was home to it. Such ecosystem-based cultural diversity was vital, fostering ways of sustainable living in different parts of Earth. Today the ecological languages of Aboriginal people, and the cultural diversity they represent, are as endangered as tropical forest species and for the same reasons: the world is being homogenized, ecosystems are being simplified, diversity is declining, variety is being lost. Ecocentric ethics challenges today's economic globalization that ignores the ecological wisdom embedded in diverse cultures, and destroys them for short-term profit.

Principle 6: Ecocentric Ethics Support Social Justice

Many of the injustices within human society hinge on inequality. As such they comprise a subset of the larger injustices and inequities visited by humans on Earth's ecosystems and their species. With its extended forms of community, ecocentrism emphasizes the importance of all interactive components of Earth, including many whose functions are largely unknown. Thus the intrinsic value of all ecosystem parts, organic and inorganic, is established without prohibiting their careful use. "Diversity with equality" is the standard—an ecological law based on Nature's functioning that provides an ethical guideline for human society.

Social ecologists justly criticize the hierarchical organization within cultures that discriminates against the powerless, especially against disadvantaged women and children. The argument that progress toward sustainable living will be impeded until cultural advancement eases the tensions arising from social injustice and gender inequality, is correct as far as it goes. What it fails to consider is the current rapid degradation of Earth's ecosystems that increases inter-human tensions while foreclosing possibilities for sustainable living and for the elimination of poverty. Social justice issues, however important, cannot be resolved unless the hemorrhaging of ecosystems is stopped by putting an end to homocentric philosophies and activities.

Action Principles

Principle 7: Defend and Preserve Earth's Creative Potential

The originating powers of the Ecosphere are expressed through its resilient geographic ecosystems. Therefore, as first priority, the ecocentric philosophy urges preservation and restoration of natural ecosystems and their component species. Barring planet-destroying collisions with comets and asteroids, Earth's evolving inventiveness will continue for millions of years, hampered only where humans have destroyed whole ecosystems by exterminating species or by toxifying sediments, water and air. The permanent darkness of extinction removes strands in the organic web, reducing the beauty of the Earth and the potential for the future emergence of unique ecosystems with companion organisms, some possibly of greater-than-human sensitivity and intelligence.

"The first rule of intelligent tinkering is to save all the parts" (Aldo Leopold—*Sand County Almanac*). Actions that unmake the stability and health of the Ecosphere and its ecosystems need to be identified and publicly condemned. Among the most destructive of human activities are militarism and its gross expenditures, the mining of toxic materials, the manufacture of biological poisons in all forms, industrial farming, industrial fishing, and industrial forestry. Unless curbed, lethal technologies such as these, justified as necessary for protecting specific human populations, enriching special corporate interests, and satisfying human wants rather than needs, will lead to ever greater ecological and social disasters.

Principle 8: Reduce Human Population Size

A primary cause of ecosystem destruction and species extinctions is the burgeoning human population that already far exceeds ecologically sustainable levels. Total world population, now at 6.5 billion, is inexorably climbing by seventy-five million a year. Every additional human is an environmental "user" on a planet whose capacity to provide for all its creatures is size-limited. In all lands the pressure of numbers continues to undermine the integrity and generative functioning of terrestrial, fresh water, and marine ecosystems. Our human monoculture is overwhelming and destroying Nature's polycultures. Country by country, world population size must be reduced by reducing conceptions.

Ecocentric ethics that value Earth and its evolved systems over species, condemns the social acceptance of unlimited human fecundity. Present need to reduce numbers is greatest in wealthy countries where per capita use of energy and Earth materials is highest. A reasonable objective is the reduction to population levels as they were before the widespread use of fossil fuels; that is, to one billion or less. This will be accomplished either by intelligent policies or inevitably by plague, famine, and warfare.

Principle 9: Reduce Human Consumption of Earth Parts

The chief threat to the Ecosphere's diversity, beauty and stability is the ever increasing appropriation of the planet's goods for exclusive human uses. Such appropriation and overuse, often justified by population overgrowth, steals the livelihood of other organisms. The selfish homocentric view that humans have the right to all ecosystem components—air, land, water, organisms—is morally reprehensible. Unlike plants, we humans are "heterotrophs" (other-feeders) and must kill to feed, clothe and shelter ourselves, but this is no licence to plunder and exterminate. The accelerating consumption of Earth's vital parts is a recipe for destruction of ecodiversity and biodiversity. Wealthy nations armed with powerful technology are the chief offenders, best able to reduce consumption and share with those whose living standards are lowest, but no nation is blameless.

The eternal growth ideology of the market must be renounced, as well as the perverse industrial and economic policies based on it. The limits to growth thesis is wise. One rational step toward curbing exploitive economic expansion is the ending of public subsidies to those industries that pollute air, land or water and/or destroy organisms and soils. A philosophy of symbiosis, of living compliantly as a member of Earth's communities, will ensure the restoration of productive ecosystems. For sustainable economies, the guiding beacons are qualitative, not quantitative. "Guard the health, beauty and permanence of land, water, and air, and productivity will look after itself" (E.F. Schumacher - *Small is Beautiful*).

Principle 10: Promote Ecocentric Governance

Homocentric concepts of governance that encourage over-exploitation and destruction of Earth's ecosystems must be replaced by those beneficial to the survival and integrity of the Ecosphere and its components. Advocates for the vital structures and functions of the Ecosphere are needed as influential members of governing bodies. Such "ecopoliticians," knowledgeable about the processes of Earth and about human ecology, will give voice to the voiceless. In present centres of power, "Who speaks for wolf?" and "Who speaks for temperate rain forest?" Such questions have more than metaphorical significance—they reveal the necessity of legally safeguarding the many essential non-human components of the Ecosphere.

A body of environmental law that confers legal standing on the Ecosphere's vital structures and functions is required. Country by country, ecologically responsible people must be elected or appointed to governing bodies. Appropriate attorney-guardians will act as defendants when ecosystems and their fundamental processes are threatened. Issues will be settled on the basis of preserving ecosystem integrity, not on preserving economic gain. Over time, new bodies of law, policy, and administration will emerge as embodiments of the ecocentric philosophy, ushering in ecocentric methods of governance. Implementation will necessarily be step by slow step over the long term, as people test practical ways to represent and secure the welfare of essential, other-than-human parts of Earth and its ecosystems.

Principle 11: Spread the Message

Those who agree with the preceding principles have a duty to spread the word by education and leadership. The initial urgent task is to awaken all people to their functional dependence on Earth's ecosystems, as well as to their bonds with other species. An outward shift in focus from homocentrism to ecocentrism follows, providing an external ethical regulator for the human enterprise. Such a shift signals what must be done to perpetuate the evolutionary potential of a beautiful Ecosphere. It reveals the necessity of participating in Earth-wise community activities, each playing a personal art in sustaining the marvellous surrounding reality.

This Ecocentric manifesto is not anti-human, though it rejects chauvinistic homocentrism. By promoting a quest for abiding values—a culture of compliance and symbiosis with this lone living planet—it fosters a unifying outlook. The opposite perspective, looking inward without comprehension of the outward, is ever a danger as warring humanistic ideologies, religions, and sects clearly show. Spreading the ecological message, emphasizing humanity's shared outer reality, opens a new and promising path toward international understanding, cooperation, stability and peace.

Acknowledgements

We thank the following persons for offering critical remarks and commentaries on earlier drafts of this article: Ian Whyte, Jon Legg, Sheila Thomson, Stan Errett, Howard Clifford, Tony Cassils, Marc Saner, Steve Kurtz and Doug Woodard of Ontario; Michelle Church of Manitoba; Don Kerr and Eli Bornstein of Saskatchewan; David Orton of Nova Scotia; Alan Drengson, Bob Barrigar and Robert Harrington of British Columbia; Cathy Ripley of Alberta; Holmes Rolston III of Colorado; David Rothenberg of Massachusetts; Burton Barnes of Michigan; Paul Mosquin of North Carolina; Edward Goldsmith, Patrick Curry and Sandy Irvine of the uk, and Ariel Salleh of Australia. Their helpful reviews do not imply endorsement of this Manifesto for which the authors take full responsibility.

AFTERWORD

BY TED MOSQUIN

It is a privilege and honour to write an Afterword to this important collection of Stan Rowe's recent writings. Stan was a man with extraordinary ecological knowledge and insight about the structure, functioning and ethical value of ecosystems and nature. While he was modest, kind and cheerful, he was also an intellectual giant who ranks among the few serious original thinkers and writers in ecology and environmental philosophy. From his ecological vantage point he offered the reader a wide range of articles including critical reviews of the writings of others. The most recent of these are presented in this book from the perspective of the emerging ecocentric worldview which exposes the causes of the broken Earth/human relationship. The originality and strength of Stan's writings is that they inevitably place us humans in our ecological context: "We are earthlings" says Stan and, as a consequence, each of us personally has a responsibility, indeed a duty, to act ethically toward Earth, its natural ecosystems and their species. But what exactly does this mean? The ecocentric worldview points the way.

I also want to reflect on what Stan meant to me as a friend and teacher, and here I feel that I speak for many others whom I know have been profoundly influenced by his writings and teachings. I am grateful to Stan's good friend Don Kerr, who, back in the early '90s convinced Stan to assemble some of his most important speeches, articles and essays from the late 1980s and get them published in the volume *Home Place: Essays on Ecology*. And once again Don assisted Stan in assembling and editing a selection of Stan's most recent writings (1992–2004) which the reader will find in this new book. From these collections of texts readers can come to comprehend the ecological foundations behind the ecocentric world view and develop personal visions for commitment and action. The two books are natural companions, belonging together on everyone's bookshelf.

I first met Stan Rowe in the winter of 1953–54 at the University of Manitoba where he was a graduate student in forestry and I had entered my second year in what was then the Science Department. Stan, some fourteen years my elder had already lived through WWII as a conscientious objector, had experienced a jail term and had taught a number of subjects to interned Japanese Canadians in a detention camp in New Denver, BC—

a part of his life that he describes in this book. Stan knew something about the ways of the bigger world while my limited experience was that of an enthusiastic young naturalist from an isolated small farm in eastern Manitoba but with an insatiable curiosity about the ways of Nature. I soon found out that Stan was a kindred spirit and we would talk about the Manitoba flora using Latin names. A penchant for remembering the Latin names of hundreds of plant species was something we had in common.

Stan's graduate research was supported by the Canadian Forestry Service, and as fortune would have it, I landed a summer job as a field assistant to Stan in his thesis research project. Forest site classification theory was all the rage among foresters of the day and Stan was searching for 'indicator species' whose presence would aid in classification of different boreal forest sites.

Under Stan's gentle guidance and tutelage I spent one of the most memorable summers of my youth. From early May to August's end, along with two other undergraduates, we camped out, living in canvas tents, cooking on wood fires in far-flung wild areas of Riding Mountain National Park, the Duck Mountains and Porcupine Hills of western Manitoba, and the boreal forests of northern Saskatchewan. We studied soil profiles, identified plants, collected plant vouchers, made notes on canopy densities and growth rates in a wide variety of coniferous, mixed and deciduous forests, and, of course, swatted mosquitoes and blackflies. Stan maintained a strict camp protocol for organizational and safety reasons. The very first task in setting up a new camp was the delegation of 'latrine duty' where the three students took turns with spade, hammer, nails and rope in building a simple latrine which we would use for the duration of our stay, often for as long as a week. Stan had his own tent and I particularly recall him getting out of his sleeping bag each morning, tiny mirror hung on aspen trunk, and shaving while joyously reciting sonnets by Wordsworth, poems of Byron and others—all from memory. Music and song were always important to Stan and that summer he brought along his clarinet and would play beautiful melodies on quiet evenings. He was gifted for his ability to sing in perfect harmony . . . which is the way he knew the Earth . . . as a participant in a living symphony of evolving Nature. That fall, after returning to university, I bought myself a clarinet.

Stan introduced me to the taxonomy of mosses, which form the vital lowest carpet layer of the ground cover in boreal forests. He revelled in slowly enunciating some of the beautiful Latin names of mosses, two of which are poetry in themselves: *Hylocomium proliferatum* and *Calliergon schreiberi*, names that still resonate through my mind after these many years.

During the 1960s and '70s we went our own ways but as research scientists in forestry and botany respectively our paths inevitably crossed when giving presentations on our work at scientific meetings of the Canadian Botanical Association, the American Institute of Biological Sciences, the Ecological Society of America and at International Botanical Congresses.

In 1961, Stan wrote a pivotal, now classical paper in ecology, entitled "The Level of Integration Concept and Ecology" published in the journal, *Ecology*. This reflected his capacity to reflect upon and grasp the three-dimensional ways in which ecosystems were structured, organized and integrated. The article formed a foundation for much of his later writings (and that of others) in the study of ecosystems as well as in Stan's work in developing some powerful ecologically founded ideas in environmental philosophy and ecological ethics—topics so well covered in his writings. The 1961 article continues to be much cited to this day.

Employed by the Canadian Forestry Service in the 1960s, Stan had an opportunity to travel the country, and in 1972 wrote the most widely used and cited book in Canadian forestry: *The Forest Regions of Canada*, a book which synthesized the knowledge on forest ecology of the day. In it he proposed a classification for Canadian forests still used today. He soon became a sought-after popular speaker at forestry conferences, environmental organization meetings and an advisor to governments. In the '70s and early '80s Stan's professional life reads like a whirlwind of invited lectures, speeches, book chapters and critiques of forest management policies. There was a multitude of essays, all emanating from his understanding of the integrity and evolutionary dynamics of geo-ecosystems and landscape ecology. From his holistic ecological insights, so well expressed in writings and speeches, his advice was sought by governments respecting problems of environment, agriculture, land-use planning and uses of technology. Inevitably, his knowledge of the structure and functions of ecosystems set the framework for presentations and writings, and in the last ten years of his life he often remarked that he was able to write from different perspectives because he could cannibalize the foundational basic ideas about ecocentric philosophy from his previous writings.

A gradual but momentous turning point in Stan's life and thought came in the mid 1980s when he turned his attention from ecological research to environmental philosophy and particularly to the application of his knowledge and insights of the workings of ecosystems to the emerging field of ecological ethics. He was then moving to the very vanguard of new thinking about the place of humans in relation to the Earth. The title of this book (*Earth Alive*) reflects the radical but honest and truthful conclusions he began to draw in the '80s, namely that Earth and its generative qualities were of profound and supreme value and that dependent humans had no idea what they were slowly destroying when they (falsely) considered themselves to be the centre and pinnacle of all value. Fortunately for me, I had been developing a growing interest in Stan's ecological writings and began to collect hard copies of texts as he produced them—two, three or more per year. As things turned out, collecting his varied texts was a good thing because by his own admission, Stan did not usually keep copies of his various essays and articles.

I share with Don Kerr, as he describes in the Preface, a memory of a pivotal time when I underwent a transformative realization that as a professional biologist and botanist I had been extremely naive about the nature and essence of my own field of knowledge. My time came in the spring of 1987. I was then working on a Carolinian forest monitoring project for Parks Canada in Point Pelee National Park and I had taken along a folder of Stan Rowe's recent essays. Two in particular I read and re-read more than a few times with ever-greater interest, and soon Stan's ecological and ethical insights began to sink in. The two are aptly titled: "This is Your Mother Calling"[1] and an unpublished essay entitled "Ecosphere Thinking."[2] Prior to that year, I had simply assumed, as homocentric scientists, and one-eyed biologists are so wont to do, that humans and science were of greatest value and that the beautiful evolved world of Nature, while deserving protection and respect was simply there, but of far lesser importance than us obviously prodigal humans. Few beliefs could be further from the truth. Stan's worldview, based on evidence from evolution and ecology, provided the rationales for recognizing the prior ethical importance of Earth and its ecosystems. Stan's profound contribution to environmental philosophy was that he was able to enunciate a worldview which placed humanity in its truthful evolutionary and ecological context. This view is in direct opposition to long held ideas found in the world's literature on ethics, which chauvinistically and dangerously assumes that ethical considerability and value only applied to the chosen human species.

In 1995, after the publication of *Home Place*, I asked Stan if he would not mind my putting up a select collection of his writings on the Internet.[3] Stan happily agreed, so with the web design help of Ray Rasmussen of Edmonton and good friend, Ian Whyte of Ottawa, the project soon got underway. More than a few of Stan's texts were written on a typewriter so I set out to digitize them, a process that can be time consuming. Soon the web site took on an evolution of its own but with Stan's writings defining its philosophical core. Peak usage is in September–October and again in January–February attesting to the use of the many texts by college students and professors.

I can also report that the last article in this book, namely "The Manifesto for Earth" was considered by Stan to be among the most important writings of his life. The Manifesto is unique among charters, proclamations, platforms and the like all of which are strongly inward looking and human-centred, or at best biocentric (when organisms are considered most important). The Manifesto stands alone because it is the first of these sorts of texts to describe and explain a thoroughly ecocentric worldview. The Manifesto finally fills a glaring conceptual gap in the ancient field of ethics where one can find thousands of books based on the narrow notion that human beings are responsible only to their fellow human beings—and for Nature too but only insofar as it is useful in serving human purposes. The Manifesto explains why we humans have duties to natural ecosystems, the Ecosphere and all Nature. Stan and I had discussed such a project since the mid-90s.

Things got serious when I visited New Denver in June of 2002 and we spent three days developing a primordial draft. We set out to write a 100 per cent ecocentric document. We agreed then that an essential and necessary element of the document was to make each principle unambiguous and as mutually exclusive as possible. Logically, how many principles might there be? A guiding question repeatedly asked was "is a principle missing?" The number kept shifting back and forth through the drafts and redrafts. Some of the reviewers' criticisms were excellent and had a profound bearing on the evolving text. The distilled eleven principles are divided into Core Principles (ecological reality, or, what is) and Action Principles (prescriptions/duties, or what ought to be). In the Action Principles, the question was: what are humanity's duties to Nature and the Ecosphere? The writing of the Manifesto was a truly joint endeavour. Had we not embarked on this joint effort, the Manifesto would certainly not have seen the light of day because Stan died only two days after seeing it published.

By way of additional explanation, in philosophical parlance the Manifesto deals with three things: *ontology*, or, what's reality? It takes the view that our immediate reality is the Earth and us in it with sunlight pouring in. Secondly, the Manifesto deals with *epistemology*, namely, how do we know reality? Again, the Manifesto says that this is by our perceptions in a science kind of way, essentially the stories as told by natural history and science. Thirdly, the Manifesto covers *ethics*, explaining what is to be valued most and how we ought to act on those values.

The Manifesto was written from an ecological, evidence-based perspective and not from a preconceived belief or faith position. Also, we as the authors saw ourselves as writing it from a distance—as if we were not participants in the processes of the Ecosphere. Rather, we visualized ourselves standing on the Moon or a far away space ship, gazing at the distant blue sphere of Earth with its swirling clouds enveloped by the blackness of space. As we looked at our home, we were mindful of the accumulated experience of two lifetimes of living on Earth as naturalists and including the experience and knowledge accumulated from two hundred years of science.

Since its appearance in the international journal 'Biodiversity'[4] the Manifesto has also been published in whole in two other publications.[5,6] A summary of the eleven principles was discussed in a recent book[7] where the Manifesto is included, and ranking it at the deepest green end of the environmental ethics spectrum (where it rightly belongs). This will further help give the Manifesto legs—something that Stan Rowe wistfully wished for a week before he died. As well, it has been translated into several languages with additional translations soon to follow.

This important book is bound to popularize Stan's powerful legacy of writings. Taken together they emphasize the supreme importance of humanity's outer shared reality. They recognize that the outer reality will forever be with us. Only a greater sense and knowledge of its importance and who we are will help to guide people along a path toward greater international understanding, cooperation and peace with Earth, all of which he recognized to be so dependent on the stability, sustaining vitality and health of the Earth's Ecosphere.

Note about the Title of this Book

Words and their precise meaning meant a lot to Stan, and many of his writings begin by explaining or defining words and phrases used in the text. Hence, it is important for the record to note that Stan's preferred title for this collection of essays was 'Earth Alive!' and not the less forceful and benign 'Earth Alive' as used in the title. To Stan, Earth Alive! reflected the importance of the scientific message being conveyed, namely the emerging realization among ecologists and evolutionists that the Earth's ecosphere itself contains the essential innate properties that enabled it to be the generative fount and creator of the stupendous variety of organisms and ecosystems over the eons. In other words, biodiversity and ecodiversity as we know them today evolved *because* the Earth's ecosphere itself contained the generative properties of aliveness. However, the publisher, NeWest Press, was faced with the important reality and practicality that, in this age of digitization, including punctuation like exclamation marks in the titles of books is problematic. If a title includes an exclamation mark, many bibliographic search engines won't locate the title unless the exclamation mark is placed in exactly the right spot. Understandably, the publisher does not wish to jeopardize the ability of booksellers and librarians to order and sell the book simply because their computer systems won't recognize the title with the exclamation mark. Hence, to reflect Stan's wishes, owners of this book may want to add a well-designed exclamation mark to the title!

References

[1] "This is Your Mother Calling" *The Briarpatch* September 1986 Posted at: http://www.ecospherics.net/pages/RoMotCal.html

[2] "Ethical Ecosphere" March 1987 (unpublished). Posted at: http://www.ecospherics.net/pages/RoEcoTh.html

[3] See http://www.ecospherics.net

[4] "The Tropical Conservancy" *Biodiversity* Vol. 5 No. 1 (2004) 3-9.

[5] *Davidsonia, a Journal of Botanical Garden Science* Vol. 15 No.2 (April 2004)70-81.

[6] Bornstein, Eli, ed. "An Ecological Ethos in Art and Architecture" *The Structurist* (43/44) (Saskatoon: University of Saskatchewan, 2003/2004) 4-9.

[7] Curry Patrick, *Ecological Ethics* (Polity Press, 2006) 173 pp.

REFERENCES

Many of these chapters are revised versions of articles, reviews or talks. They are listed below, citing the occasion on which they were first presented, in the order in which they appear in the book. In some cases, the titles been changed from the originals.

Every effort has been made to obtain permission for quoted or reprinted material. If there is an omission or error, the editor and publisher would be grateful to be so informed.

ARTICLES

Culture and Creativity: *The Structurist* No. 33/34 (1993-1994).

Traditional Ecological Knowledge: *Ecospheric Ethics Homepage:* www.ecospherics.net/pages/Ro993tek_1.html

Ecology of Cities: *The Structurist* No. 39/40 (1999/2000).

Ecology and Architecture: *The Structurist* No. 31/32 (1991-1992).

Education for the New World View: *The Trumpeter*, Vol. 8, No. 3, 1991.

What on Earth is Life? *Ecosystem Health, Vol. 7, No. 3, Sept. 2001.(Published by Blackwell Science).*

The Mechanical and the Organic: *The Structurist* No. 35/36 (1995-1996).

Earth Awareness: The Integration of Ecological, Aesthetic, and Ethical Consciousness: *The Structurist* No. 41/42 (2001/2002).

Progress is Greater Connectedness: (Originally titled Progress and Connectedness) *The Structurist* No. 37/38 (1997-1998).

REVIEWS

Rebels Against the Future: The Luddites and Their War on the Industrial Revolution; Lessons for the Computer Age by Kirkpatrick Sale (Don Mills, Ontario and New York: Addison-Wesley Publishing Company, 1996): *The Structurist* No. 35/36 (1995/1996).

How Deep Must the Change Be? Reviews of *In the Absence of the Sacred*: *The Failure of Technology and the Survival of the Indian Nations*, by Jerry Mander, (San Francisco: Sierra Club Books, 1991); and *The Dream of the Earth*, Thomas Berry (San Francisco: Sierra Club Books, 1990). (Excerpted from a larger review of three titles): *The Structurist* No. 31/32 (1991/1992).

The Ecology of Eden, Evan Eiseinberg (Toronto: Random House of Canada, 1998): *The Structurist* No. 39/40 (1999/2000).

A New Ecology? Give Me a Break! A Scathing Review of *Discordant Harmonies: A New Ecology for the 21st Century*, by Daviel B. Botkin, (New York: Oxford University Press, 1992) *The Trumpeter* Vol. 12 No. 4 (1995).

What Hopes for Ecologizing Religions. Review of *Spirit and Nature: Why the Environment is a Religious Issue*, Steven C. Rockefeller and John C. Elder, eds. (Boston: Beacon Press, 1992). (Excerpted from a larger review of five titles) *The Structurist* No. 33/34 (1993/1994).

The Resurgence of the Real: Body, Nature and Place in a Hypermodern World, by Charlene Spretinak (New York and Don Mills, Ontario: Addison-Wesley Publishing Company Inc., 1997) (Excerpted from a larger review of five titles): *The Structurist* No. 37/38 (1997/1998).

Ken Wilbur, On Transcending This Poor Earth, a review of *A Brief History of Everything*, 2000: *The Trumpeter*, Vol. 17, No. 1, 2001.

A Manifesto for Earth: *Biodiversity* Jan-March 2004, Vol. 5, Issue. 1, 3-9.

BIOGRAPHY

BY ANDREA ROWE

JOHN STANLEY ROWE, BORN JUNE 11, 1918, DIED APRIL 6, 2004

John Stanley Rowe was born in Hardisty, Alberta in 1918, the second of five children. It was probably in the United Church where his father was Minister that John (or Stan, as he was known in his professional life years later) discovered singing, a love he was to carry with him all his life. One of my fondest memories of my dad is of arriving home as a child from my dance lessons on wintry nights, and hearing his fine voice singing from the kitchen where he was preparing dinner. "Amateurish though I am at singing," he wrote in 1994, "it appeals to me more and more as an avocation, one that doesn't cause harm, confusion, or anxiety."

His greatest love was the outdoors—he took his children, and later his grandchildren, on nature hikes and pointed out the features of the plants that surrounded them. My mother called my dad the most interesting man she had ever met, and I would add that he was also the gentlest. He went through university with ease, graduating with a PhD in Botany in 1956 (University of Manitoba). Like most of his generation, the Second World

War had delayed his college years, but for dad, his gentleness surfaced when he was called to enlist. He had grown up in a pacifist, anti-war family and chose to be a conscientious objector.

While most conscientious objectors argued to be excused based on religious beliefs, dad argued from a philosophical perspective that war was morally and ethically wrong. He was sent to jail for several months—and released only after many letters were written on his behalf. Dad was later sent to New Denver, British Columbia, a beautiful town on the Slocan Lake surrounded by mountains. Here he taught science and math to the children of the interned Japanese in a one-room schoolhouse as part of his "Alternative Service."

In 1953 he married my mother, Julia Mary McQuoid, a young Registered Nurse he met at a skating party in the back yard of the rooming house in Winnipeg where they both lived. Then, after ten years in Ottawa with the Department of Forestry, he moved the family (now with son John and daughter Andrea) back out west so that he could accept an academic position in the Department of Plant Ecology in Saskatoon.

It was there that my dad's ideas about environmental ethics really took shape: "We people would not be alive without sunshine, air, water, soil, sediments, and all the millions of other marvellous organic forms—animals and plants—that together constitute the global ecological system or Ecosphere," he said at the launch of his book, Home Place. "Accepting that truth, then highest value is attached to the Ecosphere and its ecosystems as the bearers of life, rather than to organisms—yes, even including people." He would later describe himself thus: "Not a misanthrope, but a defender of Earth against the excesses of anthropes."

In 1990, he moved back to New Denver, and spent the next fourteen years reading, thinking, writing and talking about how to change people's worldview—and how to protect the physical beauty of the world around us. He never gave me much advice, but his letters often ended by encouraging me to spend as much time outdoors as I could.

It was back in New Denver that he met Katherine Chomiak, his companion for the last years of his life. After his stroke, and for the three weeks until he died, dad, Katherine and I sang songs together around his hospital bed. Although paralyzed on one side and unable to speak, he could still hum the old songs he'd always loved and when, one day, mysteriously, he did find the ability to talk to us, it was to say, simply, "I like to sing."

Andrea Rowe is Stan Rowe's daughter.

STAN ROWE: A BRIEF ACADEMIC NOTE

11 June 1918 John Stanley Rowe born in Hardisty, Alberta

1941 BSc, University of Alberta, Botany

1948 MSc, University of Nebraska, Plant Ecology

1956 PHD, University of Manitoba, Botany and Plant Ecology

1948–1967, Worked for Canadian Forestry Services, (Winnipeg and Ottawa)

1968 Professor of Plant Ecology, University of Saskatchewan

1985 Retired as Professor Emeritus, University of Saskatchewan

PUBLISHED

Forest Regions of Canada, (Ottawa: Canadian Forest Service Publication 1330) 172 pp.

Landscape Guide to the Land Forms and Ecology of Southern Saskatchewan, Environment Saskatchewan, 101 pp.

Home Place: Essays in Ecology, (Edmonton: NeWest Publishers Ltd, 1990) 253 pp.

The Sand Dunes of Lake Athabaska, by Peter Jonker and Stan Rowe, University of Saskatchewan Extension, 194 pp.

Stan said in his application to teach at the University of Saskatchewan that his "most significant theoretical contribution," mentioned by Ted Mosquin, was "The Level—of Integration Concept and Ecology," *Ecology*, 1961.

The thesis is that the community (forest stand, for example) should be replaced by the ecosystem (forest as land entity) as a first-order object-of study. . . . The community is an abstraction, but the ecosystem is objectively amenable to study from the viewpoints of morphology, ecology, physiology, and to descriptions in spatial and temporal time.

He wrote many articles on forestry, some on the north, some on landforms, the latter usually from an ecological approach. In Saskatoon he began publishing in more popular journals, *Blue Jay* and *NeWest Review* most prominently and then wrote a series of more elaborate essays, many in this book, for *The Structurist*.

He was a member of many committees, chairing the Scientific Committee for the Canadian Council of Ecological Areas, was a member of the Canadian Environmental Advisory Committee to the Federal Minister of the Environment, and many more. He was a great supporter and one of those who organized the Canadian Parks and Wilderness Society (CPAWS).

When he became Professor Emeritus he wrote in his acceptance that he'd spend more time outdoors "doing what I like to do which, fortunately, is the way I've spent my professional life."

Home Place
Essays on Ecology
by Stan Rowe
Foreword by Wes Jackson

First released in 1990, the essays in *Home Place* range from the personal—the search for a childhood vision of pristine grassland, the boy who goes from hunting to respecting wildlife and the living space around him—to theory on land use, environmental law, agriculture, education, and technology as it affects the relationships between humanity and the Ecosphere.

"Beside Rowe's unconventional, fresh intelligence, inlaid with his exacting, delightful use of words, lies a humor laced with ironic bite."
—*Saskatoon Star-Phoenix*

ISBN 10: 1-896300-53-7 / ISBN 13: 978-1-896300-53-5 PB
$24.95 CDN / $19.95 US